SHANGHAI REGENERATION | FIVE PARADIGMS
上海城市更新 | 五种策略

李翔宁　杨丁亮　黄向明　著

上海科学技术文献出版社
Shanghai Scientific and Technological Literature Press

图书在版编目（CIP）数据

上海城市更新五种策略/李翔宁，杨丁亮，黄向明著．—上海：
上海科学技术文献出版社，2017.8
　ISBN 978-7-5439-7462-3

Ⅰ.①上… Ⅱ.①李… ②杨… ③黄… Ⅲ.①城市规划—研究—上海 Ⅳ.①TU982.251

中国版本图书馆CIP数据核字（2017）第145044号

上海城市更新五种策略

责任编辑：张　树　王倍倍
助理编辑：杨怡君
出版发行：上海科学技术文献出版社
地　　址：上海市长乐路746号
经　　销：200040
经　　销：全国新华书店
印　　刷：上海望新印刷有限公司
开　　本：889×1194　1/16
印　　张：15.125
版　　次：2017年7月第1版　2017年7月1次印刷
书　　号：ISBN 978-7-5439-7462-3
定　　价：168.00元
http://www.sstlp.com

CREDITS
备注

AUTHOR/EDITOR | 作者/编者

Xiangning Li | 李翔宁

Dingliang Yang | 杨丁亮

Xiangming Huang | 黄向明

PUBLICATION DESIGN | 书籍设计

Dingliang Yang | 杨丁亮

Ruoyun Xu | 徐若云

PUBLICATION PLAN | 书籍策划

Xiangming Huang | 黄向明

Jia Fu | 傅佳

Mo Li | 李漠

PROOFREAD | 校对

Kirby Anderson | 柯比·安德森

周燕儿

SPECIAL THANKS | 特别感谢

Iñaki Ábalos | 伊纳吉·阿巴罗斯

Fionn Byrne | 菲昂·拜恩

Zhuo Cheng | 成卓

David Jimenez | 大卫·希门尼斯

Dana Kash | 达纳·卡什

Jolene Wen Hui Lee | 李文慧

Yasamin Mayyas | 雅思敏·马亚斯

Chi Yoon Min | 闵治允

Paul Chit Yan Mok | 莫哲昕

Guan Min | 闵冠

Zhe Peng | 彭哲

Wei Rong | 荣蔚

Fan Wang | 王凡

Yao Xiao | 肖瑶

Huopu Zhang | 张霍普

Leqin Zou | 邹乐勤

Bing Han | 韩冰

Xin Nie | 聂欣

Xin Wu | 吴欣

Ke Zheng | 郑科

Published by Applied Research and Design Publishing. An Imprint of ORO Editions |
上海科学技术文献出版社为国内第二版出版商

Printed in China. |
上海望新印刷厂

Copyright © 2017 by Xiangning Li, Dingliang Yang, Xiangming Huang and Tianhua Group
著作版权2017 归李翔宁，杨丁亮，黄向明和天华集团所有

All rights reserved. No part of this book may be reproduced, stored in a retrieval system, or transmitted in any form or by any means, including electronic, mechanical, photocopying of microfilming, recording, or otherwise (except that copying permitted by Sections 107 and 108 of the U.S. Copyright Law and except by reviewers for the public press) without written permission from the publisher.

版权所有。未经允许本书的任何部分不得以任何的形式（例如电子版本、物理印刷、照片影印）进行发行、传播、转载等。

The editors have attempted to acknowledge all sources of images used and apologize for any errors or omissions.

作者们已经努力确认所有使用图像的来源，如有任何错误或遗漏敬请谅解。

CONTENT

007　INTRODUCTION

008　Rethinking Shanghai: A Journey of Regeneration | Xiangning Li
010　Five Paradigms for Urban Regeneration and City Design in Shanghai | Dingliang Yang
022　Tianhua, from Architecture to Urban Practice | Xiangming Huang

046　FIVE PARADIGMS

058　Transcontextual Heterotopia: Pujiang New Town
060　Pujiang New Town: Provisional Notes on Ecological Urbanism | Fionn Byrne
082　Building the Form of a Territory | Yao Xiao with Dingliang Yang
094　On Tabula Rasa: A "Great Leap" in Urbanization | Zhuo Cheng

104　Cultural Multiplicity: Biyun International Community
106　From International Community to Global City | Ruoyun Xu
130　A Dialogue between the Local and the Global | Huopu Zhang

144　Innovative Motivation: Knowledge Innovation Community
146　Innovation District as A Glocalized Urban Strategy | Dingliang Yang, Wang Fan

182　Context Revitalization: Rainbow City
184　Rainbow City: Mixed-Use Urban Redevelopment | Yasamin Mayyas
196　Rainbow City Development Strategy: Work+Live+Play=Community? | Jolene Wen Hui Lee

212　History Reinvention: Hengmian Historical Town
214　The Unbearable Lightness of Culture | Guan Min
226　Critical Nostalgia | Paul Chit Yan Mok with Dennis Lok Kan Chau

238　BIBLIOGRAPHY

目录

007 介绍

009 再思上海：中国城市的更新之路 | 李翔宁
011 上海城市更新与设计的五种策略 | 杨丁亮
023 天华，从建筑到城区的设计实践 | 黄向明

046 五种策略

059 异质空间：浦江新镇
061 浦江新镇：景观城市主义随想 | 菲昂·拜恩
083 建设区域形态 | 肖瑶、杨丁亮
095 空白之上：城市化中的"大跃进" | 成卓

104 文化多元：碧云国际社区
107 从国际社区到全球都市 | 徐若云
131 局部与全局的对话 | 张霍普

144 创新驱动：创智天地
147 创新园区作为全球本土化的城市策略 | 杨丁亮、王凡

182 文脉再生：瑞虹新城
185 瑞虹新城：混合功能的城市再开发 | 雅思敏·马亚斯
197 瑞虹新城开发策略：工作+居住+休闲=社区？ | 李文慧

212 历史再造：横沔古镇
215 不能承受的文化之轻 | 闵冠
227 反思乡愁 | 莫哲昕、邹乐勤

239 参考引用

Xiangning Li   **Rethinking Shanghai: A Journey of Regeneration**

Following the Chicago Model of modernism and the Los Angeles Model of post-modernism, China has contributed a new mode to the world's contemporary urbanization. Similar to the Urban Sprawl represented by Los Angeles, it is of a higher density and faster speed. Chinese cities are, just like a huge sponge, absorbing the chips of European and American urban models, chewing, mixing, digesting, and pushing them to the extreme. When the urban spatial expansion reaches its limits, we will have no more unoccupied land for new constructions. Like ripples on the water, different expanding cities may also interact and collide at their boundaries. However, unlike the ripples, which run through one another, cities turn inward after the collisions, an indication that we are entering into the era of urban regeneration.

If the rapid development of Chinese cities over the past three decades, since the 1980s, needs a moment to slow down and reflect on the differences, gains, and losses in our construction modes, then we are gradually seeing the arrival of this moment. Urban regeneration is a new value judgment that enables us to rethink our built environment and review the rash misjudgments we've made in urban construction and development. It also offers us an opportunity to break the existing urban system and pattern, remodel our public space, and even our way of life. It is a new journey for us to turn inward and set sail again.

As an interdisciplinary urban idea, policy, and practice, urban regeneration covers political, economic, and cultural categories and involves sustainable development, social equality, public interest, efficiency, and many other issues. As a common problem the world is faced with in urban governance, it offers an opportunity for us to rethink the urban concept. Urban regeneration needs us to not only study the construction of urban entities but also reflect on the value system of cities in wider social and cultural perspectives.

How can an interactional social space reach organic renewal rather than incontinuous advancement? Over the past three decades, what in China's urban development is worth keeping and continuing and what needs revising and remodeling? Among all the Chinese cities, Shanghai offers the richest samples and cases. A product of Chinese urbanization, contemporary Shanghai, has faster and higher construction goals and meanwhile presents the contradiction and crisis of this development model in the most comprehensive manner. How can we study the samples from outside the system, get to the root of the issues, all while keeping a slight distance in order to reflect? This is where we can start to seek opportunities for study.

In the spring of 2016, I was invited to give a course on contemporary Chinese urban architecture at the Harvard University Graduate School of Design (Harvard GSD). During the semester's spring break, I arranged a survey of Chinese urban architecture for students registered in the course and assigned a group of them to study in depth the cases of Chinese cities. As a representative design agency, that has participated in the Chinese urbanization movement, Tianhua Architecture was also willing to join the Harvard GSD course and offer urban samples of different scales and natures.

We chose five cases. Among them, there are not only buildings and construction of international communities, but also protective remodeling of traditional Chinese towns; not only high-rise residential communities, transformed out of Shanghai's old shanty towns, but also creative communities suitable for the lifestyle of young people. Not just satellite towns built under the guidance of the concept of centralized construction ideals like 'One City, Nine Towns,' but also the continuous and piecemeal reconstruction of old areas over a dozen years.

Presently, we will conduct a critical and reflective study on these cases, trying to find the political, economic, and cultural causes behind the built environment, and analyze the gains and losses of these cases in terms of social impact, community construction, spatial quality, and architectural culture. These are carefully-chosen samples that are restricted by special conditions and limits, but they are also typical and reproducible, and thus can make useful urban regeneration references for Shanghai and even the whole country of China.

李翔宁　　**再思上海：中国城市的更新之路**

当代城市化的演进，在现代主义的芝加哥模式和后现代的洛杉矶模式之后，中国向世界贡献了快速发展的新模式，这是一种与以洛杉矶为代表的都市蔓延式发展相似但更加高密度、高速度的版本。中国城市像一块巨大的海绵，吸收着欧洲、美国城市模式的碎片，咀嚼、搅拌、消化并将其推向极致。当城市摊大饼式拓张达到极限，我们不再有更多可以新建的空白用地。如果把不同城市的拓展和相互影响看作水波纹的涟漪在彼此边界处碰撞，那么城市又和水波彼此穿越的波纹不同，只能转而向内，将前沿指向自身内部。这种转向标志着城市更新纪元的开始。

如果说二十世纪八十年代以来近三个十年的中国城市狂飙式高歌猛进的时代需要一个减速的时段并让我们可以思考我们建造的不同模式的差异或得失，那么我们已经渐渐看到这个时刻的到来。城市更新是一种新的价值判断，让我们可以重新省思我们的建成环境，检视我们在城市建设和发展中曾经作出的轻率误判，给我们一个机会打破既有城市体系和格局，重塑我们的公共空间乃至我们的生存方式。这是一个转向自身，再次启航的新旅程。

作为跨学科的城市理念、政策与实践研究，城市更新涵盖政治经济文化的众多范畴，涉及可持续、社会公平、公共利益和效率等诸多议题。它既是全球城市治理面临的共同问题，亦提供了再思考城市概念本身的机会。城市更新需要我们研究城市实体的建构，更需要我们从更广泛的社会、文化的角度再思考城市的价值体系。

一个多元互动的社会空间如何在不同的城市尺度和地域形成良性的有机更新而非断裂式的跃进？中国城市前三十年发展的哪些方面值得保存和延续，哪些方面需要修正和重塑？没有哪一座中国城市比上海更具有代表性、更能提供丰富的样本和案例。当代上海是中国式造城的产物，它更快、更高的建设目标也最全面、最深刻地呈现了这种发展模式自身的矛盾与危机。如何从系统的外部进行切片式的样本研究，既深入在地性问题的核心，又能保持一定反思的距离？这是我们寻找研究机会的出发点。

2016年春季，我受邀在哈佛大学设计研究生院讲授一门关于当代中国城市与建筑的课程。在春假的间隙要为选课的同学们安排一次中国城市建筑的调研，并有一组同学在随后的课程中针对中国城市的案例进行深入的研讨。而天华作为一个深度参与了中国式造城运动的代表性设计机构，也愿意结合哈佛设计研究生院的课程，提供不同尺度和性质的城市研究样本。

最后选定的五个研究对象，既有国际化社区的营建，也有传统中国古镇的保护性改造；既有上海老棚户区改造的高层社区，也有适合青年人群生活方式的创意社区；既有集中式建造、一城九镇概念引导下建设的卫星城，也有十几年持续零打碎敲式推进的旧区改造。

今天我们对这些案例进行批判性和反思式的研究，试图发掘建成环境背后的政治、经济、文化动因，并对样本城市案例在社会影响、社区建构、空间品质、建筑文化的诸方面得失进行深入剖析。作为切片式选取的研究样本，它们既带有各自的特殊现实条件和限定，又具有一定的代表性和可复制性，从而可以为上海乃至中国的城市更新提供有益的借鉴与参照。

Dingliang Yang

Five Paradigms for Urban Regeneration and City Design in Shanghai

Fig.1 Shanghai Skyline 1990
Fig.2 Shanghai Skyline 2010
Source: Cox, Savannah. *The Incredible Evolution of Shanghai: 1990 to 1996 to 2010.* January 21, 2014

图1　1990年上海的天际线
图2　2010年上海的天际线
来源：萨凡纳·考克斯，《上海的不可思议的演变：1990年至1996年至2010年》，2014年1月21日

Since the advent of the Open Door policy, the rise of Shanghai over the past three decades has been a noticeable exemplar of city-making. The first two decades of Shanghai's rapid urban development have been heralded by extensive new construction and the concentration of building Pudong New District, which brings the city back to the international forum and regains worldwide attention for its cityscape and architecture. *"Just as happened in the late 1920s, few cities have attracted such public fascination and media attention as Shanghai has over the past decade. The widespread interest generated by the current resurgence of China's most illustrious city is noteworthy on many levels: Shanghai's affinity historically with 'the outside world', its political and economic influence, and the sheer scale of recent regeneration have transfixed the global audience. These factors, while transcending many themes, are central to urbanism."*[1] In the most recent decade, the major developments of Shanghai, in accordance with continual urbanization, started reorienting toward the redevelopment and renovation of old city blocks, also making Shanghai deserving of global focus.

During the first two decades of Shanghai's superlative growth, 'design' acted as a strong power, shaping and reconfiguring Shanghai; this can be seen especially through the transformation of its skyline. (Fig.1-2) However, in the design realm, there is seemingly a common misunderstanding that China has only urban planning and architectural design, without urban design. Though China doesn't have a clear disciplinary definition of urban design, in the cities, especially Shanghai (a city acknowledged as the most beautiful Chinese metropolis), the design of the cityscape – from the large scale to architectural details – has always acted as one of the fundamental powers defining the city since the beginning of twentieth century in both developmental and redevelopmental stages.

After the rapid and extensive urbanization of Shanghai, the city is confronted by the new challenge of urban regeneration, though this is not the first time this has happened in the city. Mary Ninde Gamewell in her 1916 book *The Gate Way to China* described Shanghai, saying: *"Changes are going on continually all over the city. Day by day old buildings are disppearing and modern ones rising in their place."*[2] This presents the first period of Shanghai's urban redevelopment wave, after the city's own industrial transformation in the first two decades of the twentieth century. Afterwards, Shanghai experienced several other rounds of urban redevelopment. Hence, it is fair to say that for Shanghai, the issue of urban regeneration is not something new, but simply a different time facing different challenges. During the Sixth Land Planning Conference in 2014, the planning department of Shanghai, for the first time came up with the idea of *"negative growth in size of planned construction land,"*[3] which will be implemented by switching the way that land is used and in turn demanding the transformation of city development. Thus, urban regeneration has been raised to an unprecedentedly important status for Shanghai. Further, within the new master plan, Shanghai has been targeted as an attractive and outstanding global city. In order to realize this aim, Shanghai needs to pursue new developmental methods, which here we call urban regeneration of the existing urban fabric.

[1] Denison, Edward., and Ren, Guang Yu. *Building Shanghai : The Story of China's Gateway.* Chichester, England ; Hoboken, NJ: Wiley-Academy, 2006. P214.

[2] Gamewell, Mary Louise Ninde. *The Gate Way to China.* New York, Chicago: Fleming H. Revell company, 1916.

[3] Municipal People's Congress of Shanghai, Outline of the 13th Five-Year Plan for National Economic and Social Development in Shanghai. January 29, 2016.

杨丁亮　　**上海城市更新与设计的五种策略**

从改革开放起，上海过去三十多年的腾飞已经使其成为一个举世瞩目的城市建设典范。在最初的二十年里，上海高速的城市发展主要表现在扩张型的城市建设以及对于浦东新区的集中建设。（这样的建设发展）让上海重新回归到国际上的关注和讨论范围之中，也让它的城市和建筑面貌再次引起了国际关注。"*正如二十世纪二十年代晚期那样，很少有城市能像上海一样在过去的十多年间吸引到如此多的公众青睐和媒体关注。中国最突出的城市其所经历的复苏所带来的广泛影响，在很多层面值得关注：上海与海外的历史渊源，它的政治与经济的影响，以及最近令全球观众震惊的城市更新尺度。这些要素超过其他的主题，与城市主义最为息息相关。*"[1] 在最近十年的继续城市化过程中，上海的发展开始转向于关注城市更新和旧城改造，这也让其再次得到了全球的瞩目。

在上海城市化最火热的头二十年，"设计"作为一种强大的力量塑造和再塑造了上海。这一点在城市的天际线的变化中被集中地展现出来。（图1-2）然而，在设计领域，对于中国似乎有一个共同的误解：中国只有城市规划与建筑设计，并不存在城市设计。虽然从学科角度来说，中国的确尚未有明确的城市设计学科定义，但是在城市的实际进程中，"城市设计"作为塑造城市面貌的根本力量之一，自二十世纪初以来一直深深影响着中国城市发展。上海这座被认为是中国最美的大城市，无论是城市扩张还是城市更新，大到区域尺度，小到建筑的具体细节都离不开城市设计的作用。

在经历了飞速的扩张型城市化过程之后，上海正面临着内向型城市更新的新挑战。然而这并不是上海这座城市第一次应对城市更新。1916年，玛丽·宁德·加姆威尔在《中国的门户》中这样描写当时的上海，"*这座城市每一天都在经历着变化。日复一日，老的建筑在消失并被建设中的现代建筑所取代*"。[2] 其字里行间所展示的是二十世纪初期，上海在完成了自身的工业化改革之后所出现的城市更新浪潮。其后，上海还经历了多次城市更新，因此，城市更新这个议题对于上海来说并不陌生，但每一次却又面临不同的挑战。2016年进行的上海市第六次规划土地会议，首次提出了"规划建设用地规模负增长"[3]，希望通过转变土地利用方式来反推城市发展转型。就这样，城市更新被推到了前所未有的重要地位。并且，上海在新一轮城市总体规划中的愿景是一座卓越和独具魅力的"全球城市"。想要实现这个规划目标，需要寻求新的发展模式，而这一模式就是基于我们现在所提到的在已有城区进行的城市更新改造。

1 爱德华·丹尼森，广裕仁，《建设上海：中国门户的故事》，奇切斯特，英格兰；霍博肯，新泽西：威利学院，2006年，214页

2 玛丽·路易斯·宁德·加姆威尔，《中国门户》，纽约，芝加哥：弗莱明·里维尔公司，1916年

3 中共上海市委，《上海市国民经济和社会发展第十三个五年规划纲要》，2016年1月29日

Fig.3 **Transcontextual Heterotopias:** building new towns and vitalizing the suburban and exurban
Fig.4 **History Reinvention:** critical preservation, renovation and rational expansion
Fig.5 **Cultural Multiplicity:** developing semi-urban fields to international communities
Fig.6 **Innovation Motivation:** transforming district from post-industrial to innovation
Fig.7 **Context Revitalization:** reconfiguring urban enclaves to mixed-use centralities

图3 异质空间：建设新镇激活近郊和远郊
图4 历史再造：批判性保护修复与理性扩建
图5 文化多元：发展半城市化区域为国际社区
图6 创新驱动：转化后工业地块为创新创意园区
图7 文脉再生：重塑城市飞地以创造混合功能新城市中心

I have been trying to select and synthesize the design strategies that emerged during the past ten years of urban regeneration in Shanghai into five different but representative paradigms. Additionally, I aim to illustrate the power and significance of urban design through an analysis of the context of each paradigm and by narrating the design and construction processes (Fig.3-7). I will also simultaneously describe and analyze the advantages and drawbacks of each paradigm, both as a retrospective of the past and a perspective on the future.

Transcontextual Heterotopias: building new towns and revitalizing the suburban and exurban --- Around the year 2000, the total population of Shanghai was close to 20 million people, having risen from 13 million in the mid-1990s, yet individual living space was still only 13.8 square meters.[4] In order to keep increasing personal living space and improving average living conditions, as well as reduce the population density in the city's downtown, the Shanghai government announced an initiative called 'One City, Nine Towns,' aimed at bringing one million people in central Shanghai to the outskirts, while at the same time driving development there and accelerating the integration of the urban and the rural.[5] Besides easing population pressure in central Shanghai, this initiative had another invisible layer of meaning in urban regeneration, changing the land usage of the city's most desirable areas by razing many of the high-density residential areas in or near downtown, comprised of 4.281 million square meters of 'shabby and dilapidated houses,' to commercial or public land uses. For instance, one of the sites was used for the Expo in 2010, which relocated the previous onsite households to Pujiang New Town, one of the most important new towns in the initiate. Hence, this paradigm should never be comprehended purely as New Town or New City-making, but always as a duality: the redevelopment of the city center and new development of the outskirts. The latter could not happen without the former.

Since 2001, 'One City, Nine Towns' as a planning scheme had gone through a large amount of modifications; in 2006, after five years, the plan eventually went into implementation. With the intention of importing the cityscape and architectural style of specific cities or districts from a select number of western countries that trade with Shanghai, the initiative set the styles of each new town: Pujiang New Town is Italian, Songjiang New City is British, Anting New Town is German and etc. *"Designers were asked to give visual form to the spatial identity and quality of different countries, a task which in some cases was given an extremely literal interpretation at the local level by local governments and developers."*[6] This made the initiative extremely controversial, as many design proposals simply carbon copied the images of the cities from select foreign countries. One urbanist stated that *"the purpose of the initiative is to construct a*

[4] Denison, Edward., and Ren, Guang Yu. *Building Shanghai : The Story of China's Gateway*. Chichester, England; Hoboken, NJ: Wiley-Academy, 2006. P214.

[5] Zheng, Shiling. Visions on the Urban Development of Shanghai. *Shanghai new towns: searching for community and identity in a sprawling metropolis.* 010 Publishers, 2010. (Den Hartog, Harry, ed.), P4.

[6] Den Hartog, Harry, Urbanization of the Countryside. *Shanghai new towns: searching for community and identity in a sprawling metropolis*. 010 Publishers, 2010. (Den Hartog, Harry, ed.), P31.

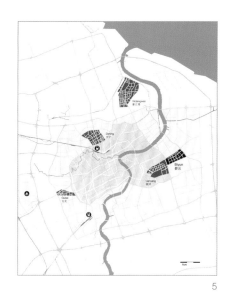
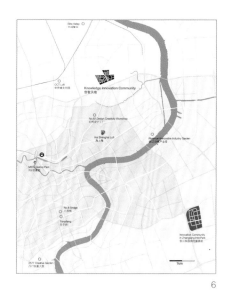
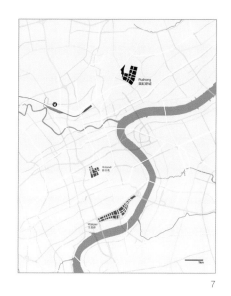

5

6

7

本文把上海在过去十年中所出现的城市更新的设计策略筛选和归纳了各具代表性的五种（图3-8），并通过分析其背景、设计和建设过程来展示城市设计对于上海这座城市的作用和意义。文章还将同时分析和描述这些策略的成功和失败之处，这既是作为一种对过去的回顾总结，也是作为一种对于未来的展望。

异质空间：建设新镇以激活近郊与远郊——到了2000年左右，上海的人口已经从二十世纪九十年代年代中期的一千三百万人口增长到了接近两千万，然而人均居住面积依然只有13.8平方米左右。[4] 为了进一步提升人均居住面积以及提高居住水平和生活质量，降低城市中心人口密度，上海政府颁布了"一城九镇"的规划战略，试图把城市中心区内一百万左右的人口迁到城郊，同时也加快城郊的发展，加速城乡一体化进程。[5] 除了疏散上海中心城人口，这一规划战略还有另一层更深的用意，就是改变上海最具价值的中心区土地的用地性质，即拆除市中心及其周边四百多万平方米的由破旧危房为主的高密度住宅区，继而将其由住宅用地转化为商业用地或公共用地。比如，其中有一块基地就涉及2010年上海世博会，被拆迁的家庭搬入了在一城九镇计划中非常重要的浦江新镇。所以对于一城九镇的理解不应该是单纯的新城新镇建设，而应该是一个二元体，作为城市中心区的更新重建与郊区新建的一种结合型的策略，而后者（新城镇建设）离不开前者（中心区更新）。

从2001年起的五年多时间里，一城九镇计划经历了多轮各种各样的调整，并于2006年开始逐步进入实施阶段。根据引进与中国有贸易往来的国外特定的不同城市和地区的建筑风格的要求，一城九镇的"风貌特色"被确定为：浦江新镇建成意大利式，松江新城建成英国风格，安亭新镇建成德国式等等。"*设计师被要求形象化空间身份和各个国家特质，但是这一任务有时被地方政府和开发商过于表面理解了。*" [6] 许多规划设计直接移植被选中的几个西方城市的"风貌"，也使之成为最大的争议。一些城市学者认为，"*一城九镇的规划目的，是用一种风格化、异域化的东西来吸引从中心往外走、能动性强、支付能力高的人。*" [7] 这样一种在郊区集合异质空间的理念不由令人想到上海城市中心区的另一种面貌，即其在租界时期，城市空间在街区尺度上拼贴混搭各国建筑特色的状态。这在一个层面上可以理解为上海一直以来是中国与西方文化的最近点，在另一个层面也

4 爱德华·丹尼森，广裕仁，《建设上海：中国门户的故事》，奇切斯特，英格兰；霍博肯，新泽西：威利学院，2006年，214页

5 郑时龄，《上海城市发展展望》，摘自哈里·邓·哈托格（编），《上海新城：追寻蔓延都市里的社区和身份》，鹿特丹：010出版社，2010年，第4页

6 哈里·邓·哈托格，《乡村城市化》，摘自哈里·邓·哈托格（编），《上海新城：追寻蔓延都市里的社区和身份》，鹿特丹：010出版社，2010年，第4页

7 刘亚晴，《上海："一城九镇"》，《那些年-瞭望东方周刊》 2015-07-09

transcontextual heterotopia through which to attract people with curiosity and financial capacity to move from the city center to the outskirts."[7] The concept of collective spatial heterogeneity on the outskirts reminds me of another face of Shanghai Center: the collective features at the urban scale of architecture from different cultures during the concession period. To this extent, Shanghai is being regarded as the most foreign city in China and also as a window into foreign cultural exhibitions. From this perspective, it is not hard to understand why Shanghai extends this type of foreignness from the center to the outskirts by theming each new town with foreign characteristics. Culturally, this paradigm integrated a trans-contextualized global metropolitan region; functionally, it constructed a large amount of residences and public facilities and introduced large-scale natural space so that a proportion of the downtown population moved to the outskirts, successfully reducing population density in the city center. However, from the design perspective, the thematized urban spaces can be seen as a reaction to a lack of identity; the paradigm becomes more a facile marketing strategy instead of an urban design methodology. Within all the new towns, Pujiang New Town is probably the only one being acknowledged as having resisted that approach. Unlike in the other towns, the architects used their professional sense and created an Italian town in spatial spirit, rather than literally mimicking the image of an Italian town.

Cultural Multiplicity: developing semi-urban fields for international communities
--- As one of the earliest treaty ports, and later the largest metropolis in the far East during the 1920s, Shanghai had a strong capacity for attracting foreign migrants because of its cosmopolitan characteristics, economic symbiosis, and cultural interactions, which were not overshadowed by the label of 'colonial.' It then became a tradition of Shanghai to be open to foreign people and cultures. Arriving in the 21st century, Shanghai has regained these charms and attractions with fast economic development and city-making. The construction of Pudong, especially Lujiazui, makes the city one of the most favored by overseas investors and visitors to China. More and more foreign immigrants are living and working in Shanghai; however, still dealing contending with language barriers. How to better accommodate the new immigrants is a challenge for Shanghai as a rejuvenated, globally-focused metropolis. A very straightforward idea is to construct international communities that can host the newcomers and help them get accustomed to Shanghai, so that later they can live in Shanghai at ease. The next question is then: where should these international communities be located?

With fast and extensive urbanization, places that were previously villages around Shanghai's old city center soon became integrated into the old town, forming new towns and supportive facilities and programs, which created a very different urban pattern, neither city nor countryside, described by John Friedmann as a 'multicentric urban field.' *"Despite its increasing urban character, the multicentric urban field still retains its rural qualities."*[8] These areas have mixed densities with an urban atmosphere, but at the same time they maintain some natural features. Also, they keep a reasonable distance from the city center. These three elements make the multicentric urban fields very good sites for international communities. Gubei Community in the north of Shanghai is the first standard international community. After 2002, Biyun and Lianyang Communities in Pudong, close to Lujiazui, were built as second-generation international communities, which brought more attention to designing better public open spaces to stimulate communication

7 Liu, Yaqing. Shanghai: One City Nine Towns. *Oriental Outlook*. 2015-07-09.

8 Friedmann, John. *China's Urban Transition*. Minneapolis: University of Minnesota Press, 2005.

可视为上海是海外文化向中国展示的一个窗口。如果从这个角度出发，也就不难理解上海想要通过一城九镇把异域混搭引申至郊区的范围，并给各新镇冠以外国特色的动因了。这种策略，在语境上整合了全球化都市区，在功效上兴建了大量的住宅和公共设施，引入了大尺度绿化，乔迁了一部分的中心区人口，降低了密度；但是就设计而言，被主题化了的城市空间呈现出对自身文化认同感的缺失，这是一种轻率的营销战略，其实不该是一种城市设计的手段。在这所有的新建城镇中，浦江新镇的设计是公认唯一相对具有抵抗性的方案，与其余所不同的是，设计师用自己的职业素养设计了一个空间精神上的意大利城镇，而非死板硬套图像上的意大利城镇。

文化多元：发展半城市化区域为国际社区——上海作为中国最早的通商口岸之一和二十世纪二十年代远东第一大都市，当时殖民的标签并没有掩盖其国际化特色，以及经济共生与文化互动的优势与特点，于是上海在历史上一直保持着对国际移民强大的吸引力。之后，对外来人群和文化的开放也成了上海的一种传统。到了二十一世纪，随着高速的经济发展和城市建设，这种吸引力在上海再度焕发。浦东新区的建设，特别是陆家嘴金融区，使其成为中国最被国外投资者或者是访客喜爱的城市。越来越多的外国人来到上海生活和工作，由于语言交流等问题，这些外国移民的生活在一定程度上受到了限制，因此上海作为一个复兴中的国际化大都市面临的挑战之一，就是如何为新来的外国移民和访客提供便利的居住和生活环境。一个直接的想法是建设国际社区，为初来乍到的外国人提供一个缓冲的区域，帮助其适应上海，为日后建立自己的生活打下基础。于是，下一个问题是：在何处建立国际社区？

随着上海的城市化急速扩张，环绕上海老城区周边的农村迅速变化，成为了融合老镇、新镇以及各种新建配套功能和设施的"既非城市亦非农村"的区域空间形态。这样的现象被弗里德曼概述为多中心的城市区域。"*尽管多中心城市区域的城市特质日益见长，但是它依然保留着原有的乡村特点。*"[8] 这些区域有着复合密度，既有城市的氛围又有乡村的自然，而且与城市中心有着适当的距离，于是就成为了国际社区很好的基地。位于城区北部的古北社区是上海第一个规模化、集约化建设的国际社区。2002年之后建立的碧云国际社区和联洋国际社区位于浦东新区，离陆家嘴不远。相比较古北而言，以碧云为代表的第二代国际社区更注重对于公共开放空间的设计，从而帮助提升居民的交流互动，在一定程度上更有助于解决语言与文化困境。2014年之后，以新江湾和大宁为代表的第三代国际社区已经开始建设。建立面向全球的国际化社区这一策略，可以被理解为通过城市更新来塑造整合型的全球都市区域。然而我们必须承认，尽管这一策略提供了高标准的生活条件，但是它引入的是一种西方的郊区化城市模式，并且在很大程度上加重了外来人群和本地人群的隔离，并没有帮助外来移民融入上海。

8 约翰·弗里德曼，《中国城市转型》，明尼阿波利斯：明尼阿波利斯大学出版社，2005

between the residents, helping them overcome the language barrier, unlike residents in Gubei Community. Since 2014, Xinjiangwan and Daning Communities became the third generation and are under construction. The paradigm of the international community can be comprehended through urban regeneration to create integrated global metropolitan regions. Knowing that this action provides better living conditions, we also have to admit that this strategy is somehow copying Western suburbanization and to a great extent creates more segregation between the foreign and the local, which is contrary to the initial design intention.

Innovation Motivation: transforming the district from post-industrial to innovation
--- After the year 2000, Shanghai was confronted with post-industrial transformation, especially for the districts of the port, production centers, and storage centers. This type of transformation related to the industrial upgrades influencing all metropolises worldwide, demanding a new form of economy based on knowledge and innovation. *"Innovation is frequently cited as the battleground of international competitiveness in the 21st century and cities are increasingly viewed as the cauldrons of innovation ... Across the globe massive renewal is taking place in our cities, fundamental shifts in the nature of work and the workplaces they host, and transformation of their output as well as their consumption."*[9] The Innovation District, a newly emerging complementary urban model worldwide, was then being introduced into Shanghai as one method of urban regeneration. As a global urban strategy, the Innovation District can be understood as a physically compact and transit-accessible area offering mixed-use housing, office, and retail in the city, anchored by leading institutions, such as universities and research centers, hospitals with extensive research and design, and companies connected with start-ups, business incubators, and accelerators.[10]

Since 2003, the local government has aligned with different developers in an attempt to duplicate the great success of urban regeneration that appeared in the Rive Gauche of Paris and the 22@ innovation district of Barcelona, or even of Silicon Valley, by taking advantage of several places with locational advantages that were previously industrial campuses that had been half-abandoned. The concept and urban strategy of the innovation district is being reinterpreted and expanded as it is introduced in Shanghai. Burgeoning innovation districts at different scales, ranging from a single building to dozens of square kilometers, have become popular urban spectacles that nurture Shanghai's transformation in different places. These districts can be found in three different types: urbanized science parks in exurban areas represented by: Zhangjiang Hi-tech Park; reimagined urban areas near or along historic Huangpu riverfronts, such as West Bund Creative Park; and the 'anchor plus' model in the downtown or midtown. Within these, the latter two are probably more successful. The most representative site of an anchor plus innovation district is the area around Wujiaochang, which matches the three "T" requirements (technology, talent, and tolerance)[11] for innovation district-making. It is

9 Leon, Nick, Attract and connect: The 22@Barcelona innovation district and the internationalisation of Barcelona business. *Innovation*. Volume 10 2008 (2-3).

10 Katz, Bruce and Wagner, Julie. The Rise of Urban Innovation Districts. *Harvard Business Review*. November 12, 2014.

11 Florida, Richard L. *The Rise of the Creative Class : And How It's Transforming Work, Leisure, Community and Everyday Life*. New York: Basic Books, 2004.

创新驱动：转化后工业地块为创新创意园区——2000年之后上海面临着后工业化的转型，特别是原先的港口区域、工业基地以及仓储基地等区域。这一类型的城市转型影响到全世界的所有大城市，它关系到产业升级且迫切地需要一种新的基于知识和创新的经济形态。"*创新常被视为21世纪的（城市或地区）国际竞争力的代表，而城市日益被视为创新的根据地……由此在全球范围内大规模的更新正在我们的城市中进行，城市中的工作性质和工作场所正在发生根本性的转变，绩效产出和消费方式也随之发生转变。*"[9] 创新创意园区作为在国际上一个新兴和饱受赞誉的城市模型被引入上海，成为城市更新的模式之一。作为一种通用的城市策略，创新创意园区（的具体概念）可以被理解为一个由领先的研究机构所引领的交通便利可达、空间紧凑、提供各种完善的住宅、办公和商业的城市区域。这里说到的领先的研究机构包括大学、研究中心或者是包含研究与设计的医院、国际大公司和创业公司的聚集地、孵化器或加速器等。[10]

自从2003年起，上海地方政府就开始与多家开发商合作，尝试在当时具有地理优势的半废弃后工业地块进行城市更新，希望在上海重现巴黎左岸、巴塞罗那22@创新区，甚至是硅谷所带来的城市更新的巨大成功。创新创意园区的概念和策略在引入上海后被重新解读、再定义，而且得到了进一步扩展，随后（创新园区）开始在不同尺度上蓬勃发展，小到一个单独的建筑，大到几十平方公里的整个区域，一同在城市的各个角落支撑起上海的转型。这些创新园区主要有三种类型：第一种是在城市远郊区的"城市化科技园"，最典型的代表就是张江高科技园区；其次是"重构的城市区域"，这些地方有着一定（工业）历史并靠近或者沿着黄浦江，比如西岸创意园；最后就是在城市中心或者分中心的"智慧+"模式的创新园区。在这三种类型中，后两者可能更为成功。最具代表性的"智慧+"模式的创新创意园区就是紧邻上海著名学府复旦大学和同济大学的上海五角场周边区域，这一地区满足创新园区建设的三个基本要求，即科学技术、人才和包容度。[11] 创智天地项目就坐落于此，（在当时的环境下）它迫切地需要一种能满足开放创新和合作共赢理念的新型空间，于是，设计师们改变了建筑乃至整个区域的空间布局与设计方式。城市设计师和建筑师通过构建一个高密度居住工作的形态，并利用公共交通的支持，建立了一个有充足的开放空间的人性化小尺度的布行街区。在此，住宅、办公和辅助设施都得到了高度的融合，从而使得整个空间具有了一种激发高效、包容和可持续经济发展的独特潜能。

9 尼克·里昂，《吸引与连接：22@巴萨罗那创新区与巴塞罗那商业国际化》，《创新》杂志，第10期，2008，2-3页

10 布鲁斯·卡兹与朱莉·瓦格纳，《城市创新区的崛起》，《哈佛商业评论》，2014年11月12日

11 理查德·佛罗里达，《创意阶层的崛起及其改变工作、休闲、社区与日常生活的方式》，纽约：基本书店，2004年

adjacent to famous institutes like Fudan University and Tongji University. This project is called Knowledge and Innovation Community (KIC), and it urges new forms of physical space under the principle of open innovation that rewards collaboration, resulting in a transformation of how buildings and entire districts are designed and spatially arrayed by designers. By presenting more dense residential and employment patterns and leveraging mass transit, urban designers and architects have created a more walkable neighborhood with sufficient open space and a smaller block size where housing, jobs, and amenities intermix, creating the unique potential to spur productive, inclusive, and sustainable economic development.

Context Revitalization: reconfiguring urban enclaves to mixed-use centralities
--- Unprecedented urbanization has reconfigured Shanghai, in both good and bad ways. Besides the acknowledged improvements to the cityscape and living conditions, there also exist some issues and challenges. The deconstruction and reconstruction of large areas of the high-density, compact city center was, for the first time, threatening the unique texture of Shanghai's traditional urban fabric – a texture formed over many decades by the amalgamation of disparate communities, built in different periods, all coming together to conduct business harmoniously. *"Vast swathes of land were leased by government, bulldozed and redeveloped for an entirely new type of clientele. The buildings erected in place of the intimacy provided by former alleyways, houses, shops, and small business, have been largely exclusive, homogeneous, unilateral developments denying public access, participation and interaction."*[12] Under these circumstances, until the 21st century, the typical urban fabric with traditional architecture from the major urban texture had become separated fragments in central Shanghai. Here we want to borrow the term "enclave"[13] to define the situation of the traditional fabric in the city center, currently in a relatively poor condition and surrounded by the modern community. The most important issue in the urban regeneration procedure during the past ten years has been how to ameliorate the built environment of existing urban enclaves and reactivate these places, making them again attractive centralities while avoiding the drawbacks on urban space brought about by previous, brutal urban redevelopment methods.

Around 2000, Shanghai Xintiandi became a project that combined the renovation strategies of classical Lilong Housing and Shikumen, integration, and maximization of open space, as well as a controlled and reasonable amount of new high-rise housing construction. It is an exemplar that has been approved to be a successful urban design and development tactic. This tactic was later borrowed and implemented in several other similar redevelopment projects in the inner ring of Shanghai. In 2010, Rainbow City appeared as another key urban regeneration project because of its scale of 1.7 million square meters and its central location next to the Bund and Suhe Creek. From the urban design perspective, compared with the previous cases, Rainbow City was also based on a more mature strategy of mixing preservation, renovation, and new development, but more importantly it managed to leverage public transportation to create a multidimensionally convenient mixed-use urban centrality.

History Reinvention: critical preservation, renovation, and rational expansion ---
Within the planned urban structure of the 1-9-6-6 model that comprises one central city, nine decentralized new cities, sixty small towns, and six hundred villages, the pace of

---

12 Denison, Edward., and Ren, Guang Yu. *Building Shanghai : The Story of China's Gateway*. Chichester, England; Hoboken, NJ: Wiley-Academy, 2006.

13 Zhou, Min. *Chinatown: The socioeconomic potential of an urban enclave*. Temple University Press, 2010.

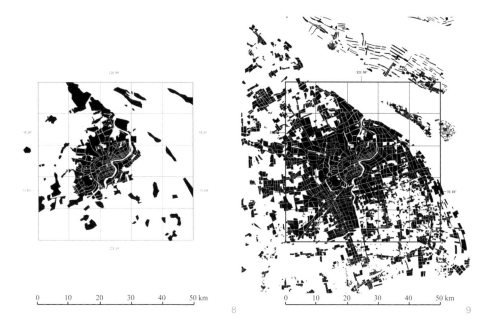

Fig.8 Urbanized Area of Shanghai in 2005
Fig.9 Urbanized Area of Shanghai in 2010
Source: Global Metropolitan Observatory, UC Berkeley

图8 2005年上海的城市化区域
图9 2010年上海的城市化区域
来源：加州大学伯克利分校，全球都会区域观测研究中心

文脉再生：重塑城市飞地以创造混合功能新城市中心——史无前例的城市化进程改变了上海，这种模式既有优势又有劣势。除了大家公认的对于市容市貌的提升和生活居住条件的改善，同样也存在一些问题和挑战。城市中心区大片的高密度、低层高的老区被拆除重建，这也是上海传统的城市肌理首次受到威胁。这种肌理是近百年演变而成的，是不同时期建造的不同社区长期并置、组合、共生以及合理运作所形成的结果。"*大量的土地被政府转让出去，然后被推土机推平，全部重建成千篇一律的（商业住宅小区）类型。于是历史积淀下来的人性化的亲切的街巷、住宅、商店和小型商业空间形态在很大程度上已经被拒绝公共进入、参与或互动的冰冷而同质化的新建建筑在原位取代。*"[12] 在这样的情况下，到了21世纪，对于上海的老城区来说，传统的建筑形态和城市肌理已经从原先城市的主体变成了市中心零落分散的片段。我们可以借用"飞地"[13]的概念来形容现在在城市中心被各种现代形式的小区包围着的那些相对破败的传统肌理社区。在城市更新过程中如何避免之前那种粗放式的城市更新给城市的空间形态所带来的弊端，但是又要合理地改善现有城市飞地的环境和面貌，并且把它们重新激活成具有高度吸引力的城市中心地带，这是过去十年城市中心范围内城市更新最为重要的议题之一。

在2000年前后，上海新天地作为一个结合了保护翻新石库门里弄等传统建筑、整合梳理公共开放空间再伴以适度的高层住宅社区开发的综合城市更新策略，为之后的中心区改进更新提供了一个很好的范例。这一城市设计和开发策略在上海被证明是成功的，而且在随后的一些内环范围内类似的更新项目中也被借鉴和采用。2010年，瑞虹新城作为又一个上海中心区的旧城综合更新项目，以其一百七十万平方米的庞大规模和靠近外滩及苏州河的地理位置成为最近几年又一个至关重要的城市更新项目。在城市设计上，相较于之前的案例，除了保护、改造和新建综合的基本策略，瑞虹新城同时最大化地利用了公共交通，意在塑造多维度的、便利的新城市综合中心区。

历史再造： 批判性保护修复与理性扩建——在"1-9-6-6"的城镇结构体系的基础上（包括1个中心城市，9个分散化新中心城，60个小城镇和600个中心村），

---

12 爱德华·丹尼森，广裕仁，《建设上海：中国门户的故事》，奇切斯特，英格兰；霍博肯，新泽西：威利学院，2006

13 周敏，《中国城：城市飞地的社会经济学潜力》，天普大学出版社，2010

urbanization has accelerated. It is dependent on the development of rail transportation and roads, which makes the border of the urbanized area of Shanghai continue to expand, resulting in its absorption of more and more historic districts, ancient towns, and villages (Fig.8-9). During this progress, different historic estates were faced with various situations that generally could be categorized into three types: first is the small-scale historical urban tissue located in downtown Shanghai, of which only a few spaces with high cultural value were preserved and renovated, then planted with new commercial programs inside, such as the area of Jing'an Temple and Yuyuan Garden; second are the historic towns previously in the suburbs, that are now becoming subcenters or even centers of the city, experiencing large amounts of demolition and new construction, while at the same time still keeping a certain amount of traditional neighborhoods and relics (for instance, Nanxiang Historic Town); finally, the historic villages on the far outskirts of the city, most of which were preserved as a whole before becoming transformed through new construction into tourism destinations and whose townscapes are treated as historical and cultural resources. Zhujiajiao Town is one of the most representative of this third type. In the last decade, when mentioning historic estates, even in the context of these three very different types, we were actually discussing the same issue: the conflict and coexistence of historic culture and memory, and the seemingly contradictory new forms of business. Even the topic of nostalgia, an overwhelmingly popular topic in Shanghai and China as a whole, is globally the same discussion: historical towns are encountering the opportunities and challenges brought by new business models.

When Disneyland entered Shanghai, the historic towns and villages around it faced unprecedented commercial opportunity, but also the biggest urban questions: should they be preserved or erased for new construction? Should they implement the strategy of critical preservation, renovation, and rational expansion? Rem Koolhaas used to express confusion about historic preservation, saying: *"what exactly needs to be preserved? And how?"*[14] There's no clear answer to this. Is it just simple conservation or pseudo-historic construction-making? In Shanghai, sadly there happen to be many superficially remade, archaized buildings. However, as Frederic Jameson criticized, it is *"its own representation of itself."*[15] Hengming Historic Town presents some new explorations in urban design, transcending outdated dialectics such as past-future, traditional-modern, and us-them. It involved no large-scale demolition and reconstruction, nor archaized buildings, but instead integrated preservation with controlled, rational new development, fighting for the vitality of environmental, social, and economic sustainability. More and more projects in similar conditions are appearing, but this design and development strategy still demands further examination to see whether it is effective or not.

Finally, to briefly summarize, the narrative of these five paradigms doesn't intend to include all the urban regeneration phenomena and strategies happening in Shanghai. Rather, the topic has selected five representative ones to indicate to the readers other types of urban redevelopment from the most recent decades Shanghai has experienced, beyond the most well-known projects like the construction of Pudong New District and the revitalization of the Bund. Within the five paradigms, there are lessons to be learned and also initiatives that deserve appreciation and promotion.

14 Koolhaas, Remmet. *Preservation of History. (Lecture).* Harvard University Graduate School of Design.

15 Jameson, Fredric. Nostalgia for the Present. *Postmodernism, Or, The Cultural Logic of Late Capitalism.* Post-contemporary Interventions. Durham: Duke University Press, 1991.

上海进一步加快了区域性的城市化进程，而且依托轨道交通和道路的发展，使城市化的边界不断向外扩展，(图8-9) 让越来越多的历史街区和古村镇进入了城市的范围之中。在这一过程中，不同的历史村镇在不同的情况下有了各自的境遇，但是大体可以理解为三类：第一，一些城市中心的历史街区、少数规模较小但是意义深远的建筑，作为文化古迹被保留了下来，并且在城市更新的过程中按照原貌进行了翻修，然后适当地加入一些新的商业内容，比如静安寺和豫园等；第二，原先地处近郊但是随着城市的扩展而地处中心或者是副中心的一些古镇，则是在大范围的拆迁和新建的同时也保留了一定量的历史街区和古迹，比如南翔古镇；第三，处于远郊的历史村镇，在通过新建转化成旅游目的地之前，大多数远效历史城镇受到整体性保护，且城镇景观被视为历史文化资源，比较有代表性的就是朱家角。过去的十年，在我们提及历史街区、村镇，即使是面对三个不同的类型，都无一例外在谈文化历史记忆，以及与它们看似冲突的当代商业形式，还有两者之间可能存在的冲突与共存。近几年在上海乃至整个中国，关于历史村镇很火热的一个话题"乡愁"，其实也并未脱离这个问题：历史古迹与新的商业形式相互冲撞所带来的机遇与困境。

所以，当迪士尼公园落户上海的时候，它周边的古镇第一次面临如此大的商业机遇，但同时也第一次面临如此大的挑战：该不该保留？又如何在这样的城市更新前提下进行批判性保护，同时理性地拓展新的开发？面对这样的议题，库哈斯曾表示："*历史保护有一定的困惑，什么样的东西需要被保护？怎么保护？*"[14] 这并没有一个明确标准。是简单地保存，还是仿古地建造？在上海，存在着大量肤浅复制的仿古建筑，但是正如詹明信的评论，"它们终究只是对于（历史）的另一种展示形式。"[15] 所以横沔代表的迪士尼公园周边的古镇做出了新的探索：在城市设计策略上，超越了简单的过去与未来、传统与现代、你与我之间的二元对立，试图去融合环境、社会、历史和经济各个方面的可持续性，不是大拆大建也不是盲目仿古，而是做到批判性的保护更新和理性的开发相结合。这一类型的城市更新项目正在越来越多地出现，这样的设计开发策略究竟成功与否仍有待时间的检验。

最后，作为简单的总结，这五种策略并非试图囊括上海在过去十年所出现的城市更新与设计的全部现象和方法，更多的是通过选取和剖析五种重要而具有代表性的策略，来向大家展示除了举世瞩目的浦东新区和外滩沿岸的建设之外，上海最近十年所经历的城市更新历程。其中有值得吸取的经验教训，也有值得推崇和推广的成功举措。

14 雷姆·库哈斯，《历史的保护》（讲座），哈佛大学设计研究生院

15 弗雷德里克·詹明信，《当下的怀旧》，摘自《后现代主义，或，资本主义晚期的文化逻辑（后当代介入）》，达勒姆：杜克大学出版社，1991

Xiangming Huang

Tianhua
from Architecture to Urban Practice

Tianhua, as one of the earliest private design agencies in China, was established on the eve of the outbreak of the extensive urbanization of the country in 1997. Our exploration at the beginning gradually turned our focus to the reflection on our own living environment and its connection with our city. Soon after the establishment of Tianhua, the wave of urbanization that is sweeping the entire country, pushed Tianhua to the forefront and expended the architects' practice from architecture to community, and further to the fragments of city. Today, after twenty years, we found our works have left subtle impact on the city of Shanghai. To some extent, it is an extension of its development from a port city one hundred years ago to a modernized metropolitan. Thus our evaluation of our works should start with the modern history of Shanghai as a background.

Here we selected five projects in Shanghai which reflect different scales, timings, also differences and representativeness caused by the intertwined forces of developers and regional governments. After putting them together on one map, you may find them seemingly different like a collage, which exactly reflected the rapid development of the city in the past twenty years. However, if we take a more in-depth observation, we will have a far more important discovery about the essential similarity under the influence of government policies and social mechanism.

During the urbanism and urban regeneration in the past twenty years, the use of "differential land" has become a very important approach. Both the fundemental development of land and the renovation of built areas are based on the introduction of residence with higher paying capability, thus the "Gentrification" is considered an inevitable result. To encourage new residents to pay more, the developer intended to highlight the difference instead of the integration between the newly-built project and the original environment. It was also very difficult to mix people with urban functions. Meanwhile, the gentrification caused ignorance on public transportation and supporting facilities. Over time, different areas of the city have been fragmented and become difficult to coordinate with each other. Those are shared problems for the development of China's cities in recent years and important background for studying our cases.

Perhaps it is too early to comment today on the development of the city in the past ten years as many of the social, economic and lifestyle effects have not yet emerged. Just as when we took over the design of KIC Phase II project, there was barely anything along the Daxue Avenue. Our design is based on the reflection on its previous failures. It had been flourishing and become a successful model of urban community life before our design was completed. This rate of change was quite common in Chinese cities in the past few years, which brought us great challenges on designing and planning. Therefore we tried to work with a team from Harvard GSD to review and reflect our previous works from the perspective of a third party, which is particularly meaningful to us.

Pujiang New Town

Pujiang New Town is located in the southern district of Shanghai. Starting from the city center, one can drive along the North-South Elevated Road across the Huangpu River through Lupu Bridge, then cross the area of former 2010 Shanghai Expo and get on the Outer Ring Road to Pujiang New Town. Initially you will be greeted by a grassy slope neatly extending along the roadside with rows of trees on top of them. A boulevard cuts the slope and leads you to a clean and neat district. The first one along the boulevard

黄向明

天华
从建筑到城区的设计实践

天华，作为最早的民营设计机构，诞生于中国大规模城市化爆发前夜的1997年。最初的摸索使得天华的建筑师们渐渐聚焦在对于自身生存和生活环境的思考，以及与其所在的城市的联系上。而在天华成立不久之后，席卷全国的城市化巨浪把天华推向前沿，也使得建筑师们主动抑或被动地把实践的尺度迅速地从建筑延展到社区，乃至城市的片断。二十年后的今天，我们再来审视这些工作，它们对上海这个城市的变化产生了微妙的影响。而且在某种程度上，是从一百多年前上海作为一个开埠城市，走上了现代化道路之后，城市发展的一种延续。对于这些工作的考察和评价，也应该以上海城市的近代史作为背景来展开。

在这里我们选择了五个上海的案例，它们反映了不同的尺度、时间性，开发主体和地区政府交织在一起的合力所产生的差异性和代表性。把这些案例放在同一张地图来看，它们似乎各不相同，像是被拼贴在一起，这也许正好反映了过去二十年中城市变化发生的速度。但如果我们进行深入的观察，更为重要的发现却是，在政策和机制的作用下所带来的实质上的相似性。

在过去二十年的城市化和城市更新的过程中，"土地级差"的利用是一个十分重要的手段。土地的基础开发和既有建成区的改造，都是以引入更高支付能力的居民为前提，这样看来"士绅化"就是其必然的结果。为了让新的居民有更高的支付意愿，开发者有意突出新建项目与原有环境的差距，而非与其融合。人群和城市功能的混合也非常困难。同时"士绅化"也使得忽视公共交通和城市配套成为可能。久而久之，很多城市区域相互割裂，很难相互协同。这些成为中国近年来城市发展中的一些通病，也是考察我们的实践案例的重要背景。

也许今天来评价过去二十年的城市发展还为时尚早，很多在社会、经济和生活方式层面的影响还没有显现出来。正如我们接手进行创智天地二期项目的设计时，大学路还十分凋零，我们的设计是基于对它的失败进行反思。然而在我们的设计还没有建成的时候，大学路却已经兴盛起来，成为城市街道生活成功的典范。这种变化的速度在过去几年的中国城市中很常见，给规划和设计带来了很大的挑战。也正是这样，我们与哈佛GSD的一个小组来共同审视及反思，希望通过第三方的视角，重新观察和认识我们曾经做过的工作，这显得格外有意义。

浦江新镇

浦江新镇位于上海市区的南部，从市中心出发，沿南北高架路从卢浦大桥越过黄浦江，穿越2010年上海世博会旧址的区域，再经过外环线，就抵达了浦江新镇。最初映入眼帘的是路边整齐的绿色草坡，草坡顶上种植着一排排的绿树。绿坡出现了一段缺口，一条林荫大道引入到一个干净整齐的城区。沿路第一幢是一个二层的建筑物。U字型布局，中间半围合成一个小广场。面对广场的柱廊中间矗立着一座塔楼，塔的端部有一平台向西延伸，可以瞭望整个新镇。西面的部分是一

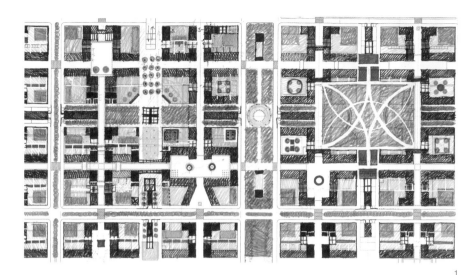

Fig.1 Sketch of Typical Cluster Plan in Pujiang New Town

图1 浦江新镇典型建筑组团平面图手绘

is a two-story building of U-shaped layout, outlining a small plaza. A tower that stands in between the colonnades against the plaza has a terrace stretching to the west on the roof, where one can lookout onto the whole town. The western part of the area is covered by a pergolas in grid with a peaceful pool and simple lawn below. It is the most important public building in Pujiang New Town – Sino-Italy Cultural Center. The huge red beams and stone pillars, cornices and walls give unparalleled features to the building, making it a primary location to host many different art exhibitions and cultural events. It is also the office and sales center of the developer. Behind the cultural center, there are cannels, streets, and well-placed buildings hidden behind the trees.

Pujiang New Town was part of the strategic urbanization of Shanghai since the beginning of the 21st century. With the rapid economic development, large numbers of people swarm into Shanghai, the core areas of the city became very crowded and the urban fringe kept expanding, which caused huge pressure on residence, transportation, environment and social services. Confronting this situation, one important countermeasure is to build a series of new towns at the the periphery of Shanghai to relieve the overcrowding of the downtown area and to stimulate the integrated development of both suburban and central districts of Shanghai. This choice itself is beyond reproach, what surprised us is the government's strategy to build these ten new towns, namely "One City and Nine Towns", based on different urban typologies of different areas in different countries. The starting point is to embrace diversity in the large-scale construction, yet it is still a bit bizarre to see the British-style Thames Town, the Anting German Town and many other towns of foreign style around Shanghai. Later, when Vittorio Gregotti and Augusto Canard, our partners on this project asked us why the government decided to do so, I explained that China, throughout history, has recognized itself as the center of the world. The Chinese garden is reflective of the Chinese understanding of itself within the universe – a garden surrounded by miniature mountains, rivers, lakes, and oceans. There is a spacial experience that occurs between heaven and earth. Realizing there are other countries in the world, we naturally use this analogy as an attitude in our national culture, similar to the Old Summer Palace built by Express Cixi in Qing Dynasty. It is understandable especially when the country is reviving after one hundred years of declining, and the reasonableness is totally a different matter.

Pujiang New Town was designated for Italian style, which meant the next step was to find the perfect urban plan for this new town; and in the competition of urban planning concept, Gregotti Associati Internation from Milan, won against the other architect, from America. Gregotti was then selected to design the urban plan for Pujiang New Town. We joined later as co-designer and local architect to adjust and develop their concept. We also collaborated with Shanghai Urban Planning and Design Research Institute to edit the regulatory document on detailed planning and urban design.

In the design competition of Pujiang New Town urban planning, the most surprising part was that the judges selected a rationalistic concept among three proposals (Fig. 1-2), which is completely different from other new town plans that were mainly based on style and features. Vittorio Gregottii and Augusto Cagnardi put their design in this way: the nature of Italy in this design is relected via the experience of thousand years of urban

Fig.2 Sketch of Typical Residential Buildings in Pujiang New Town

图2 浦江新镇典型居住建筑手绘

个田字型的廊架覆盖的空间，在它的下面有一个平静的水池和一片简朴的草坪，这是浦江新镇最重要的公共建筑——中意文化中心。巨大的红色过梁和石材的柱子，檐口和墙体，一起给了这个建筑物无与伦比的特性。在这里经常有各类艺术展览和文化活动，同时也是开发商在新城办公和销售的场所。经过了这栋建筑之后，是运河、街道以及掩映在绿树后面错落有致的各种建筑。

浦江新镇的缘起是21世纪初上海市深入推动城市化的一项策略。随着经济的高速发展，大量人口持续涌入上海，核心城区非常拥挤，城市边缘不断蔓延。居住、交通、环境以及社会服务等面临巨大的压力。针对这种状况，上海市的一项重要对策是在核心城区的外围建造一连串的新城新镇，以此来解决城区内过度拥挤的问题，并且以此带动上海周边城郊地区与城市中心的一体化发展。这种选择本身并没有什么值得非议的地方，比较出乎大家意料的是政府为这十座新城新镇，即"一城九镇"，选择了以不同国家地区的城市建筑作为其风貌原型的策略。虽然其出发点是要解决在大规模建设的条件下如何应对多样性的问题，但在上海周边出现英式风格的泰晤士小镇、德式安亭小镇等等，还是让人有些匪夷所思。后来，当这个项目的合作者维多里奥·格雷高蒂和奥古斯托·卡尼亚迪问我们为什么要这么做的问题时，我试图这样来解释，中国历史上自认为是世界的中心，中国的园林体现了中国式的宇宙观，被缩小的山、河、湖、海围绕四周，以实现天地之间的空间体验。而当我们知道世界上还有其他国家之后，就把这种自然模拟演变成一种民族文化的微缩版本，就像以前清朝的慈禧太后在圆明园中所做的一样。特别是中国积弱百年后，重新崛起之际，这种思维并不是不能理解的，至于是不是合乎情理那就另当别论了。

浦江新镇被定为意大利城镇。于是为这个新城找到一个完美的城市规划的工作就如火如荼地展开了。最终总体概念规划设计竞赛由来自米兰的格雷高蒂事务所获胜，其方案被选为浦江新镇的实施方案。我们作为合作设计师和当地建筑师随后加入，对概念方案进行调整深化，并与上海城市规划设计研究院一起，合作展开控制性详细规划和城市设计的编制工作。

浦江新镇总体规划的设计竞赛中，最令人惊奇之处是评委在三个方案中选择了一个理性主义的规划方案（图1-2），使之与其他的新城规划从风貌出发的做法完全不同。格雷高蒂和卡尼亚迪是这样论述这个设计的：意大利性在这个设计里体

life in this country. The basic components are: city center, the definition of boundaries, key public buildings and spaces, networks of roads and rationality of transportation, pedestrian system, city parks, river networks throughout the city, riverbank life and possible water transportation. The complicated relations between thme is presented in a very straightforward way. First, we built a green slope along the main road to the city center, sectioning the new town from surrounding districts like a city wall, while protecting the town from the bustling on the streets. The center of the new town is extended along the central river, connecting the Puxing Road and Huangpu River as well as the water transportation and the core area of Shanghai. Linear to this, is a place facilitated with commercial buildings, offices, municipal depots, cultural centers, universities, etc. The road network of 300m*300m became the transportation grid of the whole town. The 600m*600m road network connects the new town to outer areas while the 300m*300m mainly includes internal traffic routes. Each 300m*300m stands for a basic block unit and is divided into four small plots by two perpendicular green belts. The plots are further connected by the pedestrain system. The buildings, arrayed along the pedestrian roads, formed the blocks of the town and hide the parking and green land behind.

Such utopian blueprint was challenged at the very beginning. The existing developable parcel was not expanding along the center of the new town to the surrounding areas. Instead, it was originally a residential land started from the town's entrance at north end. In order to adapt to the changes in the plan, we need to create a place for public facilities in the north entrance of the new town, acting as the center for this area, so as to activate new residential buildings and public events that result in pulling the entire community together. Therefore, the Sino-Italy Cultural Center, as we mentioned above, was designed as a public facility to spread the vision of the new town and to promote cultural exchanges with Italy. The residential facilities at the very beginning included high-rise apartments, townhouses and several waterfront villas. (Fig.3-4) The designs for the townhouses were interesting because each house was L-shaped, the outer side contributed to the boundaries along the streets, and the inside was a fenced-up yard. The enclosed yard is actually commonly seen in both traditional Italian and Chinese residential buildings. Said yard was added to the defining of the streets and urban spaces, which was able to avoid the suburbanization that American residential communities have. High-rise residential towers were placed at the two ends of the green belt as land marks, same as the towers in some Italian cities.

Pujiang New Town, with more than fifteen years of development, has taken shape by now. However, many details in our planning are not yet figured out, and there are a lot of things that are turning out to be completely different from what was expected at the beginning. The city center was not built the way we planned, which is our biggest disappointment. Therefore, Pujiang New Town is not functioning in an integrated way, more like a "sleeping" town. With the expansion of the central districts of Shanghai, it is no longer an independent town but is fused into the entire core area of the city.

Biyun International Community and Biyun Garden

Twenty years ago, Jinqiao was an industrial processing zone far from the city center of Shanghai, with no actual urban life except large factories and very few remaining farmland. The industrial land was gradually developed, and the rest is the living land that matches with it. When Mr. Yang Xiaoming, general manager of Jinqiao project

Fig.3 Waterfront Townhouses in Pujiang New Town
Fig.4 Villa Cluster in Pujiang New Town

图3 浦江新镇滨水联排别墅
图4 浦江新镇别墅区

现为意大利人千百年来城市生活的经验，其基本构成要素为：城市的中心，边界的限定，重要的公共建筑和公共空间，道路的网格和交通的合理性，步行体系和城市公园，穿越城市的河网，水岸生活和水上交通的可能性。这些复杂的关系被一种非常直白的方式展现出来。首先，我们沿着通往城市中心的主要干道筑起一道绿色的草坡，像一道城墙把城区与周围区隔开来，同时也保护城区不受干道上嘈杂繁忙的交通影响。新城的中心沿中心河道展开，连接起主要干道浦星公路、黄浦江，以及水上交通和上海的核心区。沿运河展开的中心是一个带状的线性中心，布置了商业、办公、市政中心、文化中心、大学等建筑。300米乘300米的路网成为整个新城的交通网格，其中600米乘600米是与新城外部联系的通路，300米乘300米是新城内部交通道路。每个300米乘300米的地块是一个基本的街区单元，它的中间又被横竖两条绿化带分割成四个小地块，这些小地块的中间又有步行系统联系起来。建筑物沿步行街道展开，形成城市街区。建筑的背后是停车和居民活动的绿地。

这种近乎乌托邦的蓝图从一开始就受到挑战。最初可以进行开发的地块并不是沿着新城中心向周围展开的，而是从新城北部的入口周围开始，地块也都是居住属性的地块。为了适应这种变化，就需要在新城的北入口区创造出一个重要的公共建筑和空间，使之在一段时间内成为这个片区的中心。从而避免这些新建的住宅，因为缺少有凝聚力的公共活动，而沦为荒郊中的孤岛。一个以传播这个新城的未来展望和促进与意大利的文化交流为主要功能的公共建筑出现了，就是我们一开始描述过的中意文化中心。最早的住宅包括一些高层公寓，一些联排别墅和水岸住宅。（图3-4）联排别墅的设计非常有意思，在每个别墅的地块中，建筑物都呈L型布置，它的外边成为城市街道的边界，而其他空间被围在建筑中间，形成一个庭院。这种围合内院的做法在中国和意大利的传统住宅中都不难看到，同时，它们又参与街道和城市空间的界定，从而使这些街区避免了北美社区的郊区化的情形。而高层塔楼则被安排在绿化带的两端，它们在城市的空间框架中突显出来，成为便于识别方位的标志，就像一些意大利城市中的塔楼一样。

浦江新镇从开始到现在经过了十五年的历程，一个新城已经初步形成，但是很多规划设计的部分没有得到实现，很多也与原来的设想相去甚远。其中最让人失望的是，规划中的城市中心并没有实现。这样，浦江新镇并没有完整的城市功能，而更像是一个睡城。同时，随着核心城区的进一步蔓延，新城也不再是一个独立的城区，而是融合在整个核心城区之中。

碧云国际社区和碧云花园

二十年前的金桥是一个远离上海城市中心的工业加工园区，除了大型工厂和周边所剩不多的农田之外，没有正常意义上的城市生活。但是随着工业用地逐渐开发

came to us, he was confronted with two goals: one regards profit, he hoped the new development would cover the cost of the heavy rent that was higher than the price of surrounding propertties at the time, the other one is more important: as Yang envisioned a community of internationalized lifestyle based on the development of this area. Based on this vision, he agreed to build commercial facilities, international schools, hospitals and even churches to attract foreigners who have higher incomes than local residents. Most of them were officers sent from oversea enterprises to branch offices in China, enjoying high subsidies on education and rent. Mr.Yang believed that the state-owned design companies were not able to understand or realize his vision, thus he would try with small-scale private design firms to exert stronger influence on the design. In fact, we understand the scheme we are going to put forward is not only some drawings, but also a proof of his idea. Even though it was a project with huge challenges, we felt like it was an opportunity we had been waiting for, and more importantly, Yang's goals inspired passion and creativity in us that pushed us to create an unprecedented design to the society.

First, Yang presented his wish list, in which there were two important goals: the first was to meet the habitating and living requirements of people from all over the world and to reach advanced international standards; the second was to imbed Shanghai specialties. As a native Shanghainese, Yang has a lot of nostalgia for the city. His purest architectural memory of Shanghai was from a moment sitting on the terrace of the hotel that faced the Jinjiang Hotel. Yang had introduced us to the proper scale between the neighboring units within traditional Shanghai lane house, which he firmly believed needed no more than one hundred families maximum for community togetherness, and for the purpose of it's citizens knowing one another. After that, we collected all kinds of elements we could recall of high-end housing, for example, an elevator with a view in a residential tower in Canada; the wall tiles on a luxury mansion on a hilltop in Hong Kong; top-level systems of windows and doors manufactured in Germany; we even discussed the building insulation that was rarely even seen in China. Everyone was excited; we almost saw a prototype of a high-end housing rising before our eyes. But the most important part, was how to turn the piling up of these elements, into a good design.

Jinqiao was far away from the development of periphery of the downtown area, standing in between the city and the suburb. As a suburbanized area, it was undergoing relatively intensive development. There was no urban interface, not to mention private and public spaces of good qualities. We first proposed to position this area from the perspective of urban planning, envisioning Jinqiao with a core, a center with commercials, schools, restaurants and parks. Our residential project, which situates in the core area, would respond to the standard of the city's quality with a very positive attitude. Therefore, we arrayed the towers along the streets to form an enclosed neighborhood and placed three clusters inside, each cluster hosts no more than one hundred units, and shares different entrances and exits, connecting to different streets, which borrowed the traditional living typology per the client's request. At the same time, they were connected with a yard from the inside, one end of the yard rests in the center of a cluster, where there is a pool at the highest area of the yard with water flowing through the dam and winding valleys to another cluster group, and then eventually to another pool in front of the third cluster. The pool is open to the street, which somehow changed the streetscape while allowing passers-by to have an indistinct view inside of the community that opens to a scene of beautiful gardens and trees.

完毕，剩下的就是与之配套的生活用地。当时金桥项目的总经理杨小明先生找到我们时，他面临着二重目标。其一是经济上的，他的土地是批租的，其成本已经超过了当时周边地区的房价，他希望新的开发能够覆盖这些成本而且还有利润。其二，也是更为重要的，是他设想将来基于该地区的发展，形成一个满足国际化生活方式的社区。基于这样的愿景，他愿意建造商业设施、国际学校、医院、甚至教堂来吸引收入高于当地居民的外籍居住者，他们通常都被外资企业派到上海的分支机构工作，享受高昂的教育和房租补贴。杨总认为当时的国营设计单位并不能理解和实现他的愿望，因此，他宁愿找到我们这样的小型民营设计公司，以便对设计施加更强的影响力。其实我们理解我们所要提出的方案不仅是一份建筑设计图纸，更应该是满足他心中理想的一种证明。虽然我们知道这其中面临的巨大挑战，但是这似乎是一个等待了很久的机会，更重要的是他的目标激发出我们的创造热情，做一个当时社会上没有的设计是我们共同的目标。

首先，杨总提出了他的愿望清单，最核心的莫过于以下两条：第一，要满足国际化人士的居住及生活要求，目标是达到国际先进水平；第二，要有上海特色。杨总是土生土长的上海人，有很多生活的体会。比如，他有一次和建筑师们一起坐在锦江饭店对面酒店的露台上，很肯定地说这就是他关于建筑的上海记忆。而又有一次，他让我们关注上海里弄传统居住方式中邻里单元的合适尺度，他认为一百户家庭是人与人之间还能够相互认识的极限。之后，大家都把能想到的一个高品质住宅的各种元素放在面前，比如曾经在加拿大的一座住宅楼里看到的景观电梯、香港山顶某一豪宅的外墙砖、德国生产的顶级门窗系统、甚至讨论了当时并不多见的建筑保温问题。这一切让大家非常兴奋，似乎在眼前浮现出一座高级住宅的雏形。但是更为重要的是，如何把这些元素的堆砌变为一个优秀的设计。

当时金桥远离市区周边的发展，介于城市和城郊之间。说城市是因为相对较高的开发强度，说城郊是其布局基本是城郊化模式，没有城市界面，没有高品质的私人和公共空间。基于此，我们首先提出了在规划上如何定位这个区域的想法，认为金桥要有一个核心，一个市镇中心，有商业、学校、餐馆、公园等等。而我们的项目是核心区域中的居住项目，会以积极的姿态来回应城市的品质要求。因此，我们的楼栋都沿街布置，形成围合街坊和街道空间。内部则形成三个组团，每个不超过一百户，分别有不同的出入口，与不同街道连结，这满足了业主对传统住区邻里单元的借鉴。同时，在内部又有一个花园把它们联系起来，花园的一端始于一个组团的中心，它在高台上形成一个水池，水流溢过水坝，从台上流下，通过蜿蜒的溪谷，流到另一组团，最后汇入第三个组团前的另一个池塘，这个池塘向街道打开，也让街景发生变化，同时让路人隐约看到社区内部美丽的花园和树木。

建筑物本身是一种传统的格式，似乎是在上海中心城区时常可以看到的那种。但是，建筑细节是经过仔细设计和推敲的，以便它在包容了崭新的内部使用功能之

Fig.5 Apartment Cluster of Biyun Community

图5 碧云社区公寓区街景

The building itself takes very traditional form, which is seemingly frequently seen in the downtown area of Shanghai. However, the details were carefully designed and refined, presenting an integrated quality from the traditionality in addition to brand new functions in indoor spaces. The buildings have gray sloping roofs, red-brick walls and gray-stone bases. Uncomplicated decorations, clear compositions of simple volumes, vertical lines created by columns and large-scale glass windows have all contributed to a group of residential buildings with "a sense of city." The lower floors along the street, which are occupied for F&B, formed a pleasant space for people to stop with the wooden terrace under trees. Since then, the Greencity community, as I can say, has gradually emerged with urbanized scenes of living, and the streets are getting more active and dynamic.

After this project, we also designed another residential project – Dawn Garden. (Fig.5-6) Our client is from Zhejiang, very close to Shanghai. He said he found the flavor of old Shanghai in Green City. When he went to Shanghai to visit his relatives during his childhood, the Cathay Theatre on Huaihai Road and the perfumed women in the streets are his earliest and most unforgettable memories of Shanghai. From his recollection, I felt the dynamic charm of the city, in addition to construction drawings, steel and concrete.

Besides the two projects mentioned above, we also designed villas, a hotel and the British Dulwich College within Biyun Community. We also envisioned a pedestrian system for the residents based on the existing grid of roadways. Unfortunately, we did not manage to finish that. Today were are designing our last project in Green City – Parcel S11, in which we hope our vision will be achieved in some way.

Knowledge & Innovation Community (KIC)

Knowledge & Innovation Community (KIC) is a large-scale integrated community targeted on intellectual innovation. It was jointly developed by Shui On Land and Yangpu Government fourteen years ago. It is located at Wujiaochang area in Yangpu District in northwest of Shanghai, surrounded by numerous universities and research institutions. Therefore, after cautious consideration, the district government and the developer chose to establish a novel community that combines the innovative spirit from Silicon Valley and the creative culture from the Rive Gauche to boost the transition of Yangpu District from traditional industry to intellectual economy.

The first phase of KIC includes office area in the north of Songhu Road and KIC Village in the west of Songhu Road. When Phase I was completed, the office area was very successul and attracted many well-known international sci-tech companies; while KIC Village was comparatively depressed, even University Avenue, which is busy, is noticeably quiet and cold. Shui On invited us to design Phase II of KIC and asked us to find a solution to the problems on Phase I. We believed the mixedness of functions in the west part is not enough. According to the original planning, small neighborhoods and dense road networks were finished, except for open public spaces for people to rest.

Taking these two problems into account, we made adjustments to the original urban design. Fortunately, Shui On imbeded quite diverse functions in Phase II, including the School of Management of Fudan Universtiy, a small business hotel and two office buildings for different typologies in addition to high-end residential buildings, which increased the possibility to strengthen the mixedness of this area. Based on these

Fig.6 Apartment Cluster of Biyun Community
图6 碧云社区公寓区近景

外，还呈现了一种源于传统的整体气质。灰色的坡屋顶，红砖的墙身和灰白石材的基座。没有采用繁复的装饰，而且通过简洁明了的体量组合，柱廊和大幅的玻璃窗形成的竖向线条，营造出一座具有"城市感"的住宅建筑，而它沿街底层的单元被用来作为餐饮功能使用。结合前面在树冠遮掩下的木质平台，形成街道和建筑之间的一个宜人的停留空间。可以说，从此碧云社区开始出现城市化的生活场景，街道也变得积极和活跃。

在这个项目后，我们还设计了另一个居住项目——晓园。（图5-6） 该项目的业主来自与上海相邻的浙江省。他说从碧云花园闻到了老上海的味道。他小时候来上海探望亲戚，淮海路上的国泰电影院和从他身边经过的身穿旗袍的女人留下的香水气味，是他关于城市的最早的、也是最难忘的记忆。我想，从他的言语间我们体会到了城市鲜活的魅力，富有生命，而不光是图纸、钢筋和水泥。

除了上面提到的两个项目，在碧云社区，我们还设计了一个别墅项目、一个酒店以及英国的德威学校。但是遗憾的是，我们曾经希望在原有的基于车行的道路网格上，再叠加一个适合居民步行的步行系统的设想却最终没有实现。今天，我们在设计碧云社区的最后一个项目——S11地块，我们希望我们曾经的这个想法能够以某种形式得以实现。

创智天地

创智天地是十四年前香港的瑞安房地产集团与杨浦区政府联合开发建设的、一个以知识创新为目标的大型综合社区。它位于上海西北的杨浦区五角场地区，周边有众多的大学和科研机构，因此，建设一个融合硅谷的科技创新精神和巴黎左岸的创意文化的新型社区，以此来拉动杨浦区由传统产业向智慧型经济转型，成为地区政府和开发者的一个重要选择。

创智天地的第一期包括了淞沪路以东的办公区域和淞沪路以西的居住部分——创智坊。创智天地一期完成的时候，办公部分非常成功，吸引了很多著名的跨国科技企业入驻。而路西的创智坊却比较萧条，甚至于现在非常繁华的大学路，在当时也是十分冷清。瑞安找到天华来做创智天地的二期，我们必须回答怎样来应对一期曾经有过的问题。我们认为当时的西区功能的混合度不够，原来规划中，在实现小街坊、密路网的同时，却缺少可供大家停留休憩的开放性公共空间。

针对这两个问题，我们对原有城市设计做出相应的调整。幸运的是，瑞安在创智二期中的功能安排是相当多元的。除了有高品质的住宅外，还有复旦大学的管理学院、一个小型的商务酒店以及两种不同类型的办公楼。这给我们加强这个区域混合度的努力提供了可能性。在此基础上，我们希望把这个区域的街道变得更有个性。例如，从政立路进入创智二期的入口，穿越复旦管理学院，我们希望是一

7

8

9

10

Fig.7　Aerial view of KIC Phase 2 (Rendering)
Fig.8　KIC Phase 2 Office Building Cluster
Fig.9　KIC Phase 2 Office Buildings
Fig.10　KIC Phase 2 Public Space

图7　创智天地二期鸟瞰（效果图）
图8　创智天地二期办公建筑群
图9　创智天地二期办公建筑
图10　创智天地二期公共空间

adjustments, we hoped to bring more characters to the streets of this area. For example, we wanted to place a boulevard with academic quality to pass through the School of Management of Fudan University from Zhengli Road as part of Phase II of KIC; the east-west Zhengli Road is the main road to Songhu Road as well as the traffic axis that connects the north and south areas. It also connects the management school, the hotel, the residential buildings and two office clusters. (Fig.7) The properly internalized streets between two residential clusters provided more secure and comfortable living enviroments for the residents. The roads between the two office clusters were to be connected to the surroundings of the buildings with landscapes to create a pleasant public plaza for the people who work inside the office buildings (Fig. 8-10). The plazas and street parks are imbedded in the original homogeneous networks of roads, which greatly changed the character of the community and provided a stage for diversified street life and creative culture.

Moreover, we expect this diversified and mixed community will not only be spacially colorful but will also become a lively platform for community intergrowth. Tianhua worked together with other architects in this project, for example, EMBT and Gensler. We proposed to place all public functions in the groundfloor to extend some of them, for example, F&B, services and retail from the building to the pedestrians along the streets. Meanwhile, with facilities such as the library with educational functions, fitness center, playground, conference and exhibition center that can be shared within the community, and virtual platforms offering online serives will truly help to realize an innovative community of high intelligence and sharing. Each building of Phase II is basically completed with residents gradually moving in. At the same time, the whole district is flourishing after years of economic growth. Particularly the University Avenue that used to be rather depressing is now getting busier every day, setting a very successful example for urban street life.

Although the situation is very different from what we expected at the begining, such positive changes greatly encouraged us and convinced us to activate the community by providing more integrated functions and open spaces for the public.

Rainbow City

Rainbow City is an urban regeneration project developed by Shui On Group. (Fig.11) It is located at a community named Hongzhen Old Street in the old town area in Hongkou district, and was in fact a very crowded, dilapidated shanty town. It was more than twenty years ago when I was first informed on this project. A Hong Kong developer was invited to develop this area, using Mei Foo Sun Chuen – a private housing estate of middle-class habitats in Hong Kong. At that time, I was shocked by a huge model: a base of more than ten meters long is plugged with residential buildings like a forest. It is a result of local experience from Hong Kong and imagination of China's enormous population. Fortunately, the project did not move forward, and the model remained as what it is. After a number of years, Shui On took over the project.

Shui On is also a Hong Kong based developer. However, they are more cautious on urban development. One obvious example is their project – Xintiandi, in which they remained and transplanted part of the old buildings in a densely occupied area of traditional Shanghai lane houses. New buildings were added to bring in new functions,

11

12

13

14

Fig.11　Rainbow City Aerial Current Condition
Fig.12　Rainbow City Aerial (Rendering)
Fig.13　Rainbow City Aerial Residential Cluster
Fig.14　The completed residence and public space in Rainbow City Parcel-6

图11　瑞虹新城现状鸟瞰（建设中）
图12　瑞虹新城鸟瞰（效果图）
图13　瑞虹新城部分住宅鸟瞰
图14　瑞虹新城6#地块（瑞虹璟庭）建成住宅及公共空间

条具有学院气质的林荫大道。而联通淞沪路的东西向主街政学路，是一条衔接南北二区的主要交通轴线，同时串联起学院、酒店、住宅和两个办公组团。(图7) 两个住宅组团之间的街道适度内部化，为住户提供更为安全和舒适的环境。而两个办公组团之间的道路则会与大楼周边的场地结合，并进行景观化处理，使之成为一个宜人的广场，为在办公楼里工作的人群提供一个休憩的场所（图8-10）。这样的广场和街头公园出现在原本均质的道路网格中，改变了这条街道的个性，为多样的街头生活和具有创意的文化提供了舞台。

伴随着这样的调整，我们还期望这个多样和混合的社区不但在空间上丰富多彩，而且，真正成为社区共生的一个平台。这个项目中，除了天华以外，还有诸如EMBT和Gensler这样的建筑师参与其中。但是，我们提出了一个原则，就是把所有公共属性的功能都安排在底层，这样，一些诸如餐饮、服务和零售的功能能够从建筑中延伸出来，与街道生活融合在一起。同时，很多类似学院的图书馆、健身中心、运动场以及会议展览这样的设施能够供社区共享。再加上可能的提供网上服务的虚拟平台，真正实现高度智能和共享的创新社区。创智天地二期的各栋建筑物已经基本建成，住户也正在陆续迁入。与此同时，随着经济在过去几年的蓬勃发展，整个区域已经兴旺起来了。特别值得一提的是，原本冷清的大学路也随着周边的繁荣而热闹起来，成为非常成功的城市街道生活的范本。

这虽然与我们接手这个项目时的情况大不相同，但是，这种积极的变化给我们以极大的鼓舞，使得我们对实现我们在设计中通过提供更多的复合性功能和更多的开放性公共空间来激活这片社区的期望和目标，充满了信心。

瑞虹新城

瑞虹新城是瑞安集团在上海投资开发的一个旧城改造项目。(图11) 项目位于虹口老城区一个叫做虹镇老街的地方，是一处非常拥挤、破败的棚户区。我第一次知道这个项目是二十多年前，一个香港发展商受邀投资开发这个地区，他们以香港中产阶级的居住区美孚新村作为蓝本，希望打造一百幢高层住宅，我当时被一个巨大的模型所震惊，十几米长的底盘上，密密麻麻地插满了森林般的住宅大厦。设想一下，一百幢一模一样的大楼，是什么样的景象。那是香港人以当地的经验，加上对中国庞大的人口的想象所得到的结果。好在他们后来没有进展下去，这个模型最终没有得到实现。过了若干年，瑞安接手了这个项目。

瑞安也是一家来自香港的发展商，但他们对待城市的态度却要谨慎得多。显而易见的一个案例是瑞安做的另外一个项目——新天地，他们在上海老式里弄住宅密集的地区保留和移植部分老建筑，加入一些新建筑，同时赋予新的功能，使之成为具有上海风格的娱乐休闲目的地。

turning it into an entertaining and recreational destination of typical Shanghai style. At the beginning of the project, Shui On took an approach that is very similar to Hong Kong mode – high-rising residential towers, commercials along the streets with a grand garden inside, which is basically the same to local Shanghai developers. As a Hong Kong invested developer, Shui On takes slower speed compared with domestic develoers, which is the biggest difference and exactly the reason why they are able to reevaluate and adjust their strategies when the city is growing with dramatically increased land value and pouring population. It was how we worked on Rainbow City. (Fig.12-13) The pursuit of Shui On of the quality of city and architecture, as well as its ambition on creating a lively community, has never changed through all these years.

Our first project with Shui On is on parcel No.6. Originally it was only for the use of residence with commercial uses along the streets. In design stages, Shui On decided to change the plan since it was aware of the fact that there will be two metro lines passing through the two ends of Tianhong Road, respectively set with Youdian Newly State Station and Linping Road Station. Therefore, the approximately one-mile road between the two station will welcome more passenger flow, which encouraged Shui On to change the positioning of this section of road and to turn it into an axis for the livelihood of this area by creating a series of destinations of shopping, F&B and entertainment. Thus we changed the original design of parcel 6. It was no longer a complete neighborhood: the corner was sectioned to create an internal street with one underground floor and two floors above the ground. Many shops were invited in to provide F&B, retail, children-focused facilities and clubs for the residents inside the community. At the same time, the street is connected to the residential area with an entrance for pedestrian only, offering safety and convenience to the residents. (Fig.14)

Meantime we were invited to design parcel 8, which is quite different from parcel 6. The land is super small and available for no more than two towers. The only possible space to play with is between the two towers, which somehow granted us an opportunity to experiment with different solutions for a small land. We arranged all driveways to the towers to the periphery and embedded a very peaceful yard inbetween. The yard was designed with numerous features of Chinese gardens – one will see a tranquil lake through round arch windows, a potted hardy pine tree is placed in front of the lake. There is a glass lobby at the entrance, the Chinese yard is hidden behind. Along the corridors at two sides one can reach the vestibule shared by both towers. The lower floors were cut out to invite the landscape from the yard, which offers a place for the residents to stop and stay.

Then we designed parcel 3 with American architect Benjamin Wood and his team. It is a urban-scale destination composed of many restaruants, performance spaces, entertainment facilities and a hotel, which was later integrated with parcel 10 and 6 to establish the one-mile long living axis of Ruihong area. Currently we are still working on parcel 2, 9, 1 and 7. (Fig.15) They are streching along Ruihong Road to the Peace Park at north end. Shui On and we focused on how to create the character of Ruihong Road. Facilited with shopping and F&B, it now becomes more like a leisurely avenue. Sitting at roadside cafes one can stop and watch the passers-by coming and going. Following this approach, we gradually learned the regeneration of an urban district and how the character of a community is created via its gradual formation in the past ten years. Architecture is a very important element during this process, but the biggest charm is life itself and people who spend their lives here.

Fig.15 Rainbow City Parcel-1 (Rendering)
图15 瑞虹新城1#地块（效果图）

瑞安在瑞虹新城的做法，一开始比较接近香港模式。高大的住宅楼栋，沿街布置商业，内部形成大花园。早期的项目与上海本地开发商的做法也大同小异。作为港资开发商，瑞安同中国国内的开发商相比，一个最大的差异就是开发的速度比较慢，但也正是这种较慢的开发速度，使得他们有机会在土地价值快速上升、人口不断涌入之际，能够有机会重新审视，并且对开发的思路不断进行调整。我们在瑞虹新城的工作就是这样一个过程。（图12-13）但是，瑞安通过对城市与建筑的品质的追求，努力创造一个有活力的社区的目标一直没有改变。

我们为瑞安设计的第一个项目是6号地块，这原本是一个纯住宅的地块，沿街有一些商业。但是，在设计的过程当中，瑞安方面希望改变原有的规划，因为他们了解到，天虹路的两端将有两条地铁线经过，分别设有邮电新村车站和临平路车站。这样一来，两个车站之间的这将近一公里的路，将是人们在这两个车站之间换乘时必然要经过的。这促使他们想要改变这一段路的定位，希望使之成为瑞虹地区的一条生活的轴线。具体的方法是沿着天虹路打造一连串的商业、餐饮、娱乐的目的地。于是我们对6号地块原有的设计进行修改，它不再是一个完整的住宅街坊，沿街角被切开，形成一条内街，从地下到地上二层，众多的商家将入驻这条街，为居民提供各种餐饮、零售、面向儿童的设施和小区居民的会所等。同时，在这条街上还设置一个住宅区的出入口，希望居民们能更方便地步行到这条街来，因为这里没有车辆，所以带着孩子也很安全。（图14）

与此同时，我们又得到了设计8号地块的工作。这是与6号地块十分不同的情形。地块非常小，仅能安排两幢塔楼，而唯一可以利用的空间是两幢楼之间的间距。这给了我们机会来尝试一种与其他更大的地块所不同的策略。我们把到达楼幢的车道全部布置在外围，中间设计成一个非常安静的庭院，庭院富有中国园林的气质。通过圆形的景窗可以看到一片静谧的水面，前面是一棵苍劲的盆栽松树。入口有一个玻璃的入口大厅，透过大厅就是中国庭院。顺着两边的连廊可以到达两个楼幢的门厅，楼幢的底层部分被架空，庭院的景观一直延伸进来，为住户提供休憩停留的地方。

之后，我们又与美国建筑师本杰明·伍德的事务所一起设计了3号地块。这完全是一个城市级的目的地，由众多餐厅、演艺空间、娱乐设施和酒店等组成。将来与10号地块和6号地块等一起形成瑞虹一公里的生活轴线。目前，我们还在进行2号、9号、1号、7号地块的设计，（图15）这几个地块沿着瑞虹路展开，一直延伸到最北面的和平公园。瑞安和我们最关注的是如何塑造瑞虹路的性格，路的两边还是有商业和餐饮的，但是它更像一条悠闲的林荫道，路边有咖啡馆可以坐坐，看看来来往往的路人。就是这样，在近十年的时间里，我们慢慢地理解一片城区的再生，以及在它慢慢形成的过程中，城市街道生活的个性是如何被塑造的。在这中间，建筑固然是一个重要的元素，但是生活本身以及生活在这里的人却是最大的魅力所在。

Hengmian Historic Town

Hengmian was originally one of the small watertowns around Shanghai. At the time dominated by water transportation, it was a station for commodity distribution for surrounding areas, and the streets were said to be densely crowded with thriving stores. However, following the process of modernization, especially after the traditional waterway traffic was replaced by land transportation, small towns like Hengmian were declining day by day. Recently, Hengmian has again attracted public attention as the area across the road has been developed as the first American amusement park in mainland China – the Disney Land. It greatly changed the geographical position of the small town and brought in new opportunities. However, the pressure and damage of modernization on traditional villages have been further enlarged at the same time.

Shanghai Shendi Group invited us to the design of Hengmian Project. Our work scope is limited within the expansion area of Hengmian Old Town, which means the core area of the ancient town was not included but assigned to other design teams. However, when we started our work, the team on the core area had no results for us as a basis or reference, which caused a lot of difficulties in our work. Through constant visits to the site and discussions with the client, we gradually understood the main idea of this project, which helped us to slowly sort out our entry points, ideas and strategies.

In the past two to three decades, many ancient watertowns have experienced a new wave of changes. The development of industrialization pushed the young people to leave the town and work in other cities, leaving a lot of old people behind. At the same time, immigrants from other poorer districts kept moving in, much of the architecture was renovated and amended with most of them losting their traditional context. The deficiency of employment failed both the local residents and immigrants for renovations of high quality, which made the ancient town more shabby and dilapidated. Many more ancient towns of larger scales and higher architectural values have been regenerated and re-developped, but most of them were transformed into tourist attractions. With tourists with curiosities for the history flocked in, the renovation of ancient towns were romanticized to cater to tourists. Whether it is true, it has to look like an ancient town, which does not help to protect the authenticity of history, on the contrary, the emergence of such plausible buildings flooded away the ones with real historical values.

We believe that to revitalize the ancient town, we must address the problem of its authenticity, and the best way is to revitalize the life in it. Fortunately, the surrounding areas are undergoing rapid urbanization and industrialization with the continuous development of society and economy, which make it possible to regenerate Hengmian as a living center with historical value and good services in this region. From this perspective, we proposed to retain most of the historic buildings in the ancient town and to re-organize the relations between architecture and waterfront. The relationship is also extended beyond the core area, which has become an important connection between the ancient town and the new development.

The Cross Water Street is the intersection of two rivers and used to be a focal point of Hengmian Ancient Town as it hosted major economic activities at that time. However, the relationship between the town and the waterfront has been totally abandonded today. We take it as a starting point to reform the old town and to connect the past and the future. We extended the east-west river to further west from this point and made it as one of the main skeletons of the expanded area of the ancient town. (Fig.16-17)

横沔古镇

横沔老街原本是上海周边江南水乡众多小镇中的一个。在那些以水路交通为主的年代里,它是周边地区的商品集散地。据说当时的街上商铺林立,生意兴隆。可是,随着现代化的进程,特别是传统水路交通被陆路交通取代后,像横沔老街这样的小镇便日渐衰落。近来,横沔又一次进入公众的视野,是由于与之一路之隔的区域建起了中国大陆上第一个美式游乐园——迪士尼。这极大地改变了这个小镇在原有地缘格局上的位置,给它带来了新的机会。但同时,在这里,原本已经严重的现代化对传统村镇的压力和破坏,更是被放大了好多倍。

申迪公司来找我们参与横沔的设计工作。天华这次的工作范围只是横沔古镇的扩展区。也就是说,古镇的核心部分并没有包含在内,将由另外的团队负责设计规划。而当我们开始展开工作的时候,负责古镇核心区的团队仍然没有一个工作成果可以作为我们的依据或参考,这给我们的工作带来了很多困难。但是,通过不断地对现场的走访以及与业主的交流,我们渐渐抓住了这个项目的主要脉络,我们的切入点、想法和策略也慢慢地清晰起来。

在过去的二三十年中,很多水乡古镇经历了新一波的变迁。随着工业化的发展,镇里的年轻人离乡外出生活工作,镇里留下很多老人。与此同时,来自外省更穷困地区的移民不断移入。很多建筑物被改建、翻造,但大多数已失去了传统的脉络。同时,由于没有好的工作机会,居民和移民都无力进行有品质的改造,古镇显得越发简陋、破败。有很多更具规模和建筑价值的古镇已经被重新改造开发了,但是大多数是被转化为旅游景点。游客怀着对历史的猎奇心理蜂拥而至,古镇也为了迎合游客进行浪漫化的改造。不管真的假的,总之要看起来像一个古镇。这些对历史真实性的保护没有起到好的作用,相反的,出现了众多似是而非的建筑,结果反而让有历史价值的建筑被淹没了。

我们认为,要重新振兴这些古镇必须要解决它真实性的问题,重建它的生活是最好的方法。幸运的是,随着社会经济的不断发展,周边区域城市化、产业化也非常迅猛,横沔有机会重新成为这个地区的一个有历史价值、同时又有良好服务和宜人居住环境的生活中心。从这个角度出发,我们建议在古镇部分保留大部分历史建筑,梳理建筑与水岸的关系,并在核心区之外,让这种关系得到延伸,并使其成为横沔新开发部分与古镇之间的重要联系。

十字水街是两条河道的交叉点,曾经是横沔古镇十分重要的一个结点,当年其主要的经济活动就围绕这里展开。但是,今天这种城镇和水岸的关系已经被荒废了。我们觉得要重整古镇,这个节点应该是一个最好的出发点,也是过去和未来的最好的汇聚点。依托于它,我们沿东西向的河道一直向西展开,使之成为古镇扩展区的主要骨架之一。(图16-17)

16

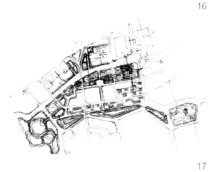
17

Fig.16 Design Sketch (Watercolor), Overall Structure of Hengmian Historic Town Renovation and Expansion
Fig.17 Design Sketch, Core Area of Hengmian Historic Town

图16 横沔古镇保护及扩张设计草图(水彩)
图17 横沔古镇保护核心区设计草图

The southwest corner of the Cross Water Street was completely ruined and was therefore not included in the scope of the core area, which enabled us to face the problems on the relationship between our workscope and the ancient town. We placed a small island extending westward. The buildings on the island have become a new center for public activities of the small town, It is connected to the other three corners of the Cross Water Street and accessible to other parts of the ancient town. The corner was historically used for storage and trade in terms of shipping in this area, which inspired us to come up with the idea to create a public plaza. Thus we designed a plaza in a ring shape with one side lifted up, and the space below is turned into a tea room. People can walk onto the plaza from the ground and look back at the ancient town or enjoy some tea time under the plaza.

The architectural volume, the scale, the diversity, the use of materials and colors of the island's waterfront are the most sensitive within the extended area, and we need to create a special interface to hide the row of middle-volume buildings behind, because new cultural and commercial functions require bigger volumes than the traditional architecture of smaller volume and scale along the waterfront, in order to respond to the diversity of the ancient town casued by smaller scales and more complicated properties due to traditional economies.

The riverbanks extending westward will embrace residential buildings of different typologies, artistic public spaces, lively schools, and parks relating to the waterfront. Thanks to the colorful treatment of the waterfront in traditional towns in south China, we kept being inspired when working on the presentations of waterfront spaces. It will become a stage for the lives in this small water town.

The north-south Mianxin Road facing the west of Hengmian Ancient Town pass through our extended area and will become the west gate of Hengmain. At the same time, it will connect residential buildings, schools, commercial facilities, community services, parks, greenland, and the remaining farmland, which will become another main axis of the extended area. Although we are concerned about the future of this broad road in the planning, it will be combined with the rail transportation under compound construction to connect this abandoned small town with the economic and social activities from other parts of the city, which anticipates an encouraging future of Hengmian.

We started from the past of Hengmian Ancient Town to study its historical context and qualities that remain through times, to further explore its connection with the future. In this course, our goal stays unchanged, which is to create a new town that embraces its history and contemporary life. We aspire to bring up a regenerated ancient town where people can travel, live and work, a community where our life will stop at eternality instead of a moment in the past.

十字水街的西南角已经完全破败了，因此，没有被划入核心区的范围，使得我们有机会在这里直面与古镇的关系问题。我们在这个角上安排一个向西延展的小岛，岛上的建筑群构成了小镇新的公共活动中心。它由石桥与原有的三个角连接，可以通往古镇的其他部分。由于历史上船运的原因，这个角以前是一片用来堆货或者交易的场地。这让我们产生了创造一个公共广场的想法。我们把这个广场做成在一头被抬起的一个环，下面的空间可以做个茶室，人们可以像上桥一样走到广场上面，在那里可以从与平时不同角度来回望这个古镇，也可以在广场的下面喝喝茶。

这个岛的水岸无论在建筑体量、尺度、多样性，还是材料及色彩的运用上，都是扩展区里面最为敏感的，以至于我们要创造一个特殊的界面来遮掩后面一排中等体量的建筑物，因为新的文化和商业功能要求比传统建筑更大的体量，而沿水的一排建筑的体量和尺度则更小，以此回应古镇由于传统经济更小的规模和更复杂的产权所造成的多样性。

向西延展的河道两岸将会出现不同类型的居住建筑、富有艺术气氛的公共空间、明快有趣的学校以及与水岸相联的公园。得益于传统江南城镇对水岸处理的多彩多姿，我们在演绎这些水岸空间时丝毫不缺乏灵感，这里将变成水乡小镇生活的舞台。

横沔古镇的西面，沔新路从北面一直往南，穿越我们的扩展区。它会成为横沔的西大门，同时，它也会串联起居住、学校、商业、社区服务以及公园、绿地和保存下来的一些农田，成为扩展区的另外一个主轴。虽然我们对这条规划中宽阔的道路感到担忧，但是，它将和复合建设的轨道交通把这个荒废的小镇和城市其他部分的经济和社会活动连接在一起。这让我们对横沔的未来感到鼓舞。

在横沔，我们虽然从古镇的过去出发，去寻找它历史的脉络和跨越时代的特质，进而探索它与未来的联系，但是，在这中间我们坚持不变的初衷，就是要创造一个既有历史的、同时也符合当代生活要求的新型城镇。我们希望通过重塑古镇，让它真正成为一个人们愿意在这里游览、生活、工作的地方，让这里的生活可以永远继续下去，而不是停留在过去的某一时刻。

Five Paradigms

The five paradigms of urban regeneration in Shanghai are defined as 1) Transcontextual Heterotopias: building new towns and vitalizing the suburban and exurban 2) Cultural Multiplicity: developing semi-urban fields for international communities 3) Innovation Motivation: transforming the district from post-industrial to innovation 4) Context Revitalization: reconfiguring urban enclaves to mixed-use centralities 5) Reinvented History: critical preservation, renovation, and rational expansion.

Each of the five paradigms is at a different scale affecting the city, from the metropolitan scale to the urban scale to the local scale, responding to different issues and requirements.

五种策略

上海城市更新的五种策略被定义为：1）异质空间：建设新镇以激活近郊与远郊 2）文化多元：发展半城市化区域为国际社区 3）创新驱动：转化后工业地块为创新创意园区 4）文脉再生：重塑城市飞地以创造混合功能新城市中心 5）历史再造：批判性保护修复与理性扩建。

这五种策略的覆盖范围从区域到中心区再到街区，各自在不同的尺度影响着城市。

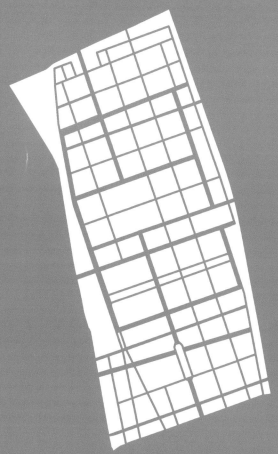

Transcontextual Heterotopias
异质空间

Pujiang New Town
浦江新镇

Building new towns and vitalizing the suburban and exurban
建设新镇以激活近郊与远郊

Pujiang New Town:

Provisional Notes on Ecological Urbanism

Fionn Byrne

The term "Landscape Urbanism" gained significant recognition in 2006 with the publication of Charles Waldheim's *The Landscape Urbanism Reader*. At that time the argument presented by Waldheim and others was that in the context of economic transformation towards a model of production and consumption that was less dependent on site, urbanization in American cities, wasn't necessarily characterized by growth, but by decline. The argument was made that a disciplinary void in the field of urbanization was open, as urban designers, developers, and architects, all understood growth, construction, and progress to be synonymous. The study of landscape was offered as an alternative, capable of rationalizing, designing, and even profiting economically and ecologically from decline. Waldheim's thesis read *"in this horizontal field of urbanization, landscape has a newfound relevance, offering a multivalent and manifold medium for the making of urban form, and in particular in the context of complex natural environments, post-industrial sites, and public infrastructure."*[1] As history would prove, the pervasive sense of decline which reads through *The Landscape Urbanism Reader* was remarkably prescient, finding even greater relevance in the context of the global financial crisis of 2007 - 2009.

Importantly, while the reordering of the urban landscape as promoted by those who would be sympathetic to the title of landscape urbanist may have been carried out in the context of economic decline, the consequent design ambition was often intended to spur future growth. The paramount example here is James Corner Field Operations' High Line, a linear park atop a defunct public infrastructure; though a post-industrial site, the site is replete with successional native vegetation. Opened in 2009 this renowned precedent project is cited as a threefold success: economically, with an increase in adjacent land values and new construction projects; culturally, as a tremendously popular destination for tourists and locals; and ecologically, by increasing urban habitat. But the High Line also stands for more than a symbol of the success of landscape urbanism. It can also be read as a shifting in attitudes from decline to growth. One could ask: if landscape is a proven medium for organizing urban form, coordinating architectural development, and establishing an underlying biophysical environment in sites characterized by economic decline, can it also provide these same services in a context of economic growth?

This question was answered with a resounding yes in Mohsen Mostafavi's *Ecological Urbanism*, published in 2010.[2] Distinguishing itself from landscape urbanism, ecological urbanism was argued to play a more projective role in an economic environment of growth. And the modifier ecological differs from landscape in a few ways. Where landscape carries an implicit tie to a physical site, ecological is a more ambiguous term associated with sustainability and allowing the concept to speak to a broader range

---

1 Waldheim, Charles. *The Landscape Urbanism Reader*. New York: Princeton Architectural Press, 2006, P15.

2 Mostafavi, Mohsen. *Ecological Urbanism*. Cambridge, Mass.: Harvard University Graduate School of Design, 2010.

菲昂·拜恩

浦江新镇：
生态城市主义随想

自2006年查尔斯·瓦德海姆《景观城市主义读本》一书面世以来，"景观城市主义"一词逐渐得到越来越多的关注。在书中，瓦德海姆与其他作者提出了以下观点：社会经济正在向"生产-消费模式"转型，逐渐不再依靠场地，在这样的时代背景下，美国的城市化进程将不再发展，反而开始衰退。这一观点的提出，意味着城市化领域开始出现学科空白，因为开发商、建筑师与城市设计师都明白，增长、建设与发展是紧密相连的。此时，景观领域研究作为一种替代性的解决方法，能合理地扭转衰退趋势、针对衰退进行设计、甚至在经济与生态层面从这种衰退中获益。瓦德海姆撰文提到："*在城市化的横向领域中，景观领域有一种新的相关性，为塑造城市形态提供了一种多价的、多形式的媒介，特别是在复杂自然环境、后工业场地以及公共基础设施的文脉中。*"[1] 就像历史已经证明的那样，《景观城市主义读本》中一直提到的无处不在的衰退是极具预见性的，这一点在2007年到2009年的全球金融危机中体现得尤为明显。

重要的是，虽然对于那些对景观城市学家这个称号有同感的人们来说，他们所推动的城市景观重组是在经济衰退时发生的，但随后的设计目标却常常被定位为未来的城市增长。最显著的例证是詹姆斯·康纳的高线公园项目，该项目是一个基于废弃公共基础设施的线性公园——实际上是一个后工业遗迹，设计者通过利用当地季节性植物来进行景观修复。自从2009年对外开放以来，这个著名案例引发了大量研究，学界普遍认为有三个层面的成功：经济上，提高了周边土地与新建建筑的价值；文化上，成为了市民与游客喜爱的公共聚集空间；生态上，增加了城市生态系统的多样性。但是，高线公园并不只是景观城市主义成功的象征，它也可以被看作是社会从衰退转向复苏的信号。有人也许要问：如果景观已经被证明是一种在经济衰退中可以组织城市形态、协调建筑开发、建立潜在生态物理环境的媒介，那么它可否在经济增长的背景下也发挥相同的作用？

这一问题在莫森·莫斯塔法维2010年的《生态城市主义》一书中得到了响亮而肯定的回答。[2] 在书中，莫森区别了生态城市主义与景观城市主义，提出让生态城市主义在经济环境增长中发挥更具针对性的作用。在莫森看来，"生态"和"景观"的区别主要有几点。第一，"景观"明确地暗示了与物理环境的紧密联系，而"生态"与物理环境的关系则较为模糊，但其与可持续性的相关性使这个概

---

[1] 查尔斯·瓦德海姆，《景观城市主义读本》，纽约：普利斯顿建筑出版社，2006年：第15页

[2] 莫森·莫斯塔法维，《生态城市主义》，坎布里奇，马塞诸萨州：哈佛大学设计学院，2010年

of design categories – technology for example. At the same time, ecological also implies a far less ambiguous and more scientific approach to site. When studied and quantified through the science of ecology, the complex natural environment recognized by landscape urbanism could be more easily measured and designed. Finally, another marked difference between landscape and ecological urbanism is the global focus of the projects in *Ecological Urbanism*. Naturally as one shifts focus from declining North American cities it becomes apparent that significant urban growth is continuing around the globe. Returning once again to *The Landscape Urbanism Reader* we read that *"Landscape Urbanism describes a disciplinary realignment currently underway in which landscape replaces architecture as the basic building block of contemporary urbanism. For many, across a range of disciplines, landscape has become both the lens through which the contemporary city is represented and the medium through which it is constructed."*[3] To which we add that environmental decline has succeeded economic decline as a driver of design, propelling ecology as a quantitative tool to understand and design the urban landscape.

Of course the history of ideas now associated with landscape urbanism extends back far beyond 2006. Indeed, as one example, many years before, Vittorio Gregotti was advocating for the importance of designing the landscape prior to individual buildings. Kenneth Frampton so notes in the foreword to *Inside Architecture*, where he says of Gregotti: *"He sees this going to ground, so to speak, as a cultural and ecological necessity, for, as he was to put it in 1983, the origins of architecture do not reside in the primitive hut but rather in the primordial marking of ground in order to delineate a human world against the unformed, chaotic indifference of the cosmos; in short, the act of culture in the void of nature."*[4] Gregotti has also had years of experience in applying his theory through the firm which bears his name. This essay will critically evaluate Gregotti Associati International's work for the Brezza Città di Pujiang, less formally known as Pujiang New Town in Shanghai, China. The project will be studied through the context of ecological urbanism, and instead of focusing on the architectural development, greater attention will be paid to the landscape design strategy. Building upon the success of the project and reflecting on some of the documented challenges, a set of provisional notes will argue for the unique and critical importance of ecology to urban design, especially in the context of rapid development as characterized by the new growth in Shanghai. In so doing, it will also be argued that considerable expertise lies in the discipline of landscape architecture which, if carefully applied, could improve future urban development.

3 Waldheim, Charles. *The Landscape Urbanism Reader*. New York: Princeton Architectural Press, 2006, P11.

4 Gregotti, Vittorio. *Inside Architecture*. Cambridge, Mass: MIT Press, 1996, XVII.

念可以与更多的设计分支互相交流，比如建筑设计与技术。第二，与"景观"相比，"生态"暗示了一种更为清晰而科学的处理场地的方式。通过生态科学的慎重而量化的方法，在景观城市主义看来十分复杂的自然环境可以更容易地被衡量与设计。第三，与景观城市主义相比，生态城市主义在项目案例分析方面更具全球视角。当人们的目光离开正在衰退的北美城市，他们自然就会发现在全球范围内城市增长还在持续进行着。回到《景观城市主义读本》，我们读到过以下论点："*景观城市主义标志着学科内部的重新洗牌，即景观正在代替建筑成为当代城市主义的基本构成单元。对于许多人来说，在众多学科中，景观已经成为再现当代城市的透镜，同时也是塑造当代城市的媒介。*"³ 在这一论点的基础之上，我们还需要加上一句：环境恶化已经代替经济衰退，成为设计的主要动力，推动生态学成为理解与设计城市景观的量化工具。

当然，远早于2006年，与如今的景观城市主义紧密相关的理论就已然出现。举例来说，事实上多年前维托里奥·格里高蒂就开始宣扬将景观设计放在单体建筑设计之前的重要性。肯尼斯·弗兰普顿在其著作《建筑之内》的前言中这样提到格里高蒂："*这样说吧，就像他1983年提出的那样，他看到了这种（重要性）作为一种文化和生态的必需，因为建筑的本源并不存在于史前棚屋里，而是存在于原始的地面标记中，其目的是为了将人类世界与宇宙中未发展的、混乱的冷漠区别开来；简单地说，（格里高蒂的所为是）在自然的空白中的文化行为。*"⁴ 格里高蒂已经通过他本人开设的建筑事务所对其理论进行了实践。本文将批判地评价格里高蒂国际设计公司在中国上海的项目"浦江清风城"，或被称为"浦江新镇"。本文将从生态城市主义文脉的角度研究该项目，重点关注其景观设计策略，而非建筑设计策略。总结并反思该项目的成功与挑战，笔者将提出一些随想，主张生态对城市设计的独特而关键的重要性，尤其是在上海城市高速发展与扩张的大环境下。通过这样的主张，本文试图论证在景观建筑学科中存在着大量的专业知识与技能，如果运用得当，将对未来的城市发展大有裨益。

3 查尔斯·瓦德海姆，《景观城市主义读本》，纽约：普利斯顿建筑出版社，2006年：第11页

4 维托里奥·格里高蒂，《建筑之内》，坎布里奇，马塞诸萨州：麻省理工学院出版社，1996年：前言，第17页

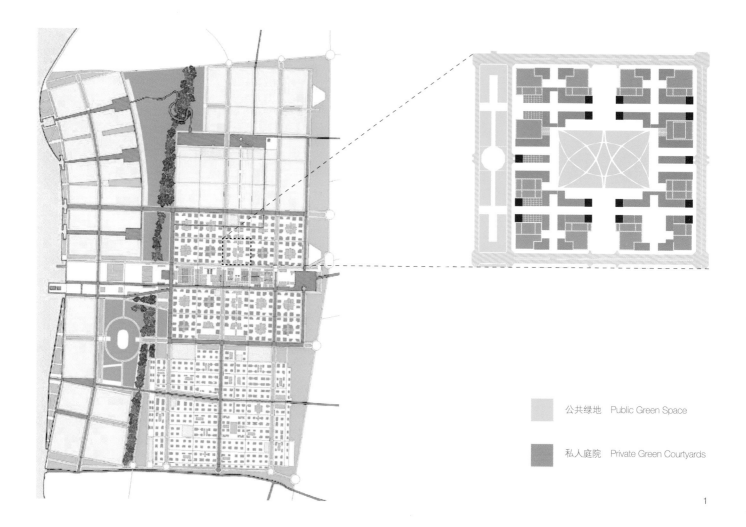

Fig.1 Master Plan of Pu Jiang New Town, and Public Green Space/Private Green Courtyard Configuration

图1 浦江新镇总平面图及街区内部绿地/庭院布局

公共绿地 Public Green Space

私人庭院 Private Green Courtyards

Pujiang New Town

In 2001 the Shanghai government proposed to build one new city and nine new towns – new cores of concentration in the greater Shanghai area, located predominantly on previously occupied farmland. This strategy of urbanization, referred to as the "principle of dispersion-concentration," was intended to increase the importance of suburban development in form and character and thus reduce pressure on the central city by attracting inner city residents to relocate to the suburbs.[5] The designers of Pujiang New Town were decided by a closed international competition, which saw Gregotti Associati International win over Scacchetti of Italy and the American firm SWA. Gregotti's plan stood out by proposing the deployment of a regular grid over the 15 km^2 site. A road network divided the site into a set of more or less equal 300 x 300 meter blocks. Each block would then be further bisected by a set of 12-meter-wide pedestrian and cycling routes running in both cardinal directions. Augusto Cagnardi, a partner in the firm, describes it in the following way: *"a framework grid was prepared in combination with strict urban rules, which according to Italian tradition guarantees architectural variety."*[6] The grid would give a consistency to the urban fabric but allow for architectural variation within the blocks. The plan was then differentiated by a central axis that would bisect the development between north and south, and would contain government buildings and other public institutions within this 300-meter-wide park. Additional local variation in the grid was provided by an existing canal system that was preserved. Finally, the perimeter of the development was bounded on three sides by a vegetated berm designed to block out the highway on the East edge and to accept future high-voltage power lines.

5 Xue, Charlie Q.L. and Zhou, Minghao. "Importation and Adaptation: building 'one city and nine towns' in Shanghai: a case study of Vittorio Gregotti's plan of Pujiang Town," *Urban Design International* 12, (2007): 22.

6 Den Hartog, Harry. "Pilot Cities and Towns: Pujiang New Town (Minhang District)," in Harry den Hartog (ed) *Shanghai New Towns*. Rotterdam: 010 Publishers, 2010, P152.

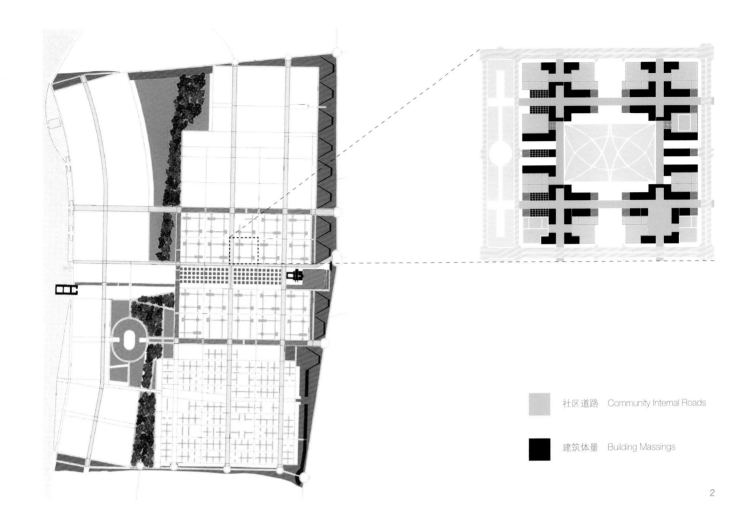

Fig.2 Master Plan of Pu Jiang New Town, and Community Internal Road / Building Massing Configuration

图2 浦江新镇总平面图及社区道路/建筑体量布局

浦江新镇

2001年，上海市政府提出计划，在之前的农业土地区域建设一个新城、九个新镇，作为大上海地区的新城市群中心。该城市化战略被称为"分散-聚集原则"，其初衷是从形式到性质全面提升城市郊区开发的重要性，吸引内城居民搬迁到城郊，减少中心城区压力。[5] 浦江新镇的最终设计者是通过一个内部国际竞赛确定的，在该竞赛中，格里高蒂国际设计公司胜过了意大利斯卡切蒂事务所与美国SWA公司。格里高蒂的方案通过将一个常规格网布局铺满15平方公里的场地而脱颖而出，其道路系统将场地划分为一系列大约300米×300米的正方形地块，每个地块被12米宽的双向行人自行车道进一步细分。（图1-2）该公司的合伙人奥古斯都·卡格纳迪这样描述这一项目："*我们提出了一套作为框架的网格，并将之与严格的城市规划政策相结合，根据意大利的传统，这将保证建筑形式的多样性。*"[6] 格网为城市肌理提供了连续性，同时也使得街区内的建筑出现各种变化。该方案还提出了一条中心轴线，将场地分为南北两部分，该轴线上是300米宽的公园，包含政府建筑与其他公共设施。另外，一条现有的河道为这一格网系统带来了变化。最后，整个场地的三条边被绿植护道所包围，将场地与高速公路分隔开，同时也为未来的高压输电线提供了空间。

5 薛求理，周明浩，《引进与适应——上海"一城九镇"建设：维托里奥·格里高蒂的浦江新镇案例分析》，《国际城市设计》，2007年第12期：第22页

6 哈里·邓·哈托格，《先锋城镇：浦江新镇（闵行区）》，摘自《上海新镇》，哈托格编著，鹿特丹：010出版社，2010年：第152页

Fig.3 Continuous Monument Project by Superstudio

图3 "连续运动"项目，超级工作室
来源：http://2.bp.blogspot.com/-yF6kJ9xKzZ8/
U6txXBDROdI/AAAAAAAAEwc/MiVT4DjNH1M/s1600/
superstudio-playa-1555x1020.jpg

Ecology is always regional

It has been remarked that Gregotti Associati's plan for Pujiang New Town began from a blank slate, a tabula rasa. This one indictment serves to negate the value of all preexisting land use. We can read that the design team could not find any preexisting historical or cultural elements to reference and as a former agricultural plain, the site of Pujiang had no significant topographic variation.[7] It was as though the site had already been prepared and devoid of culture in the manner critiqued by Kenneth Frampton. He describes the tabula rasa in disparaging terms:

"*It is self-evident that the tabula rasa tendency of modernization favors the optimum use of earth-moving equipment inasmuch as a totally flat datum is regarded as the most economic matrix upon which to predicate the rationalization of construction. Here again, one touches in concrete terms this fundamental opposition between universal civilization and autochthonous culture. The bulldozing of an irregular topography into a flat site is clearly a technocratic gesture which aspires to a condition of absolute placelessness.*"[8]

We can reference also Superstudio's well known Continuous Monument of 1969, where, even if the intentions were utopian, the project represented a planar surface as a field devoid of culture but prepared to accept a new future. The flat grid is a sign of human rationality and a contrast to nature.

It is however worth pausing here a moment to consider if a flat datum necessarily suffers from a condition of placelessness. I believe there is an important alternative argument to be made: that a lack of topography is not a lack of cultural significance. As urban development in Shanghai continues, particularly as it expands into flat agricultural land, precedent from the "prairie style" of landscape architecture, it can serve to strengthen a site's regional identity and establish a cultural continuity with the past. The prairie style was advanced in the early 20th century, most notably advocated for by the well-known landscape architects Jens Jensen and Ossian Cole Simonds. It is defined as "*an American mode of design based upon the practical needs of the middle-western people

7 Den Hartog, Harry. "Pilot Cities and Towns: Pujiang New Town (Minhang District)", in Harry den Hartog (ed) *Shanghai New Towns*. Rotterdam: 010 Publishers, 2010, P152.

8 Kenneth Frampton, "Towards a Critical Regionalism: Six Points for an Architecture of Resistance," in Hal Foster (ed) *The Anti-Aesthetic: Essays on Postmodern Culture*. New York: New Press, 1998, P26.

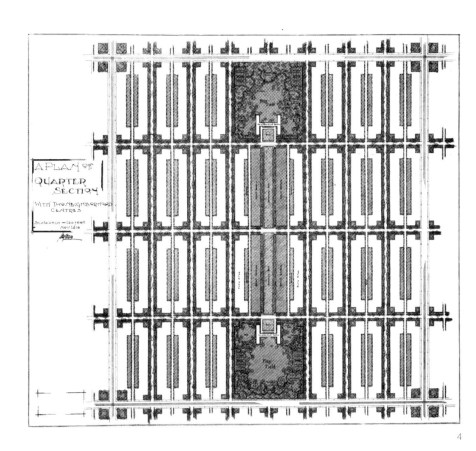

Fig.4 A Greater West Park System, a proposal for integrating parks and gardens into the city grid
Source: Jens Jensen. *A greater west park system: after the plans of Jens Jensen*. Chicago: West Chicago Park Commissioners, 1920. P44.

图4 大西部公园系统，将花园绿地融入城市格网的方案
来源：延斯·延森，《大西部公园系统：在延斯·延森的方案之后》，芝加哥，西芝加哥公园管理局，1920年．44页

生态是地域性的

一种普遍的论调是，格里高蒂国际设计公司的浦江新镇方案是从一片完全空白的场地——一张白纸——开始的。这一论断从本质上否认了原有场地的所有土地价值。可以想见，设计团队无法找到任何已有的历史或文化元素作为参考，同时，位于农业平原的场地几乎没有高程变化。[7] 这一场地看起来就像已经在物理上和文化上被清空，等待着浦江新镇的建设。肯尼斯·弗兰普顿曾经用以下的轻蔑字句对"白纸"状态进行批判：

"*不言自明地，白纸作为一种现代化潮流，对挖掘机这样的搬运土石的设备颇有青睐，因为一个完全平坦的基准面被看作是最有经济潜力的基础，它使理性化的建设成为可能。在这里，人们可以清晰地感受到在普世文明与本土文化之间的本质差异。很明显，使用推土机将非常规的高差夷为平地是一种技术统治论式的姿态，它期望一种绝对无场所的状态。*"[8]

我们还可以参考超级工作室1969年发布的著名的"连续运动方案"（图3），即使其设计意图是乌托邦式的，该方案仍然通过展现一个平坦表面，表达了试图接受新未来的文化匮乏。同时，平面网格标志着人类的理性和与自然的对立。

然而，我们应该在此稍作暂停，思考一下平面基质是否一定会带来无场所状态。我们可以提出一个重要的替代论点：缺少高程变化并不意味着缺少文化意义。随着上海城市发展的继续，尤其是当其向平坦农田扩张时，景观建筑领域的"草原风格"（图4）可以强化场地的地域特质，与场地历史建立一种文化上的连续性。草原风格兴起于20世纪早期，代表人物包括当时的著名景观建筑师延斯·延森与奥西恩·科尔·西蒙兹。草原风格被定义为"*一种美国式设计，基于美国*

[7] 哈里·邓·哈托格，《先锋城镇：浦江新镇（闵行区）》，摘自《上海新镇》，哈托格编著，鹿特丹：010出版社，2010年：第152页

[8] 肯尼斯·弗兰普顿，《迈向批判性地域主义：建筑抗御的六点提议》，摘自《反美学：后现代主义论文集》，哈尔·福斯特编著，纽约：纽约出版社，1998年，第26页

and characterized by preservation of typical western scenery, by restoration of local color, and by repetition of the horizontal line of land and sky which is the strongest feature of prairie scenery."[9] It is a distinctly regional style that, at its core, is predicated on the flat datum. What Jensen and Simonds spent considerable time advocating for in their writing and design work is that topography is only one characteristic of site, intractable from both environmental forces – weather, solar exposure, hydrology – and vegetation cover. Here, the idea is not to apply the prairie style to Pujiang, but rather to recognize that Shanghai's climate, the high water table on site, and the soil profile established through years of agriculture, will all dictate a community of vegetation best suited to the area, and will affect how local inhabitants experience place. Irrespective of architectural style, the landscape design can support a critical regionalism as so advocated for by Frampton. The rasa is itself the significant historical and cultural element of the site. A landscape design cognizant of the prairie style would seek to preserve and strengthen long and broad views, would foreground native vegetation, and would subtly manipulate landform to control microclimates; or, as more eloquently expressed: *"foreign eyes are not educated to see the slight undulations in 'flat' prairie that give so much quiet enjoyment every day to those who live on the land."* [10]

This is not a critique of Gregotti Associati's plan for Pujiang, but rather a comment on how it has been described and received. In actuality the firm has been a strong advocate for regional specificity, for what has been called '*reading the territory as an archaeological structure.*'[11] The cultural continuity, livability and environmental sustainability of rapid urbanization will be more successful if the landscape; as topography, environment, and vegetation, is firmly established within the purview of urban design – preceding architectural intervention in the same way as the framework grid in Gregotti Associati's plan for Pujiang New Town.

Ecology is both a territorial infrastructure and a precise design operation

The description of Gregotti Associati's framework grid is a strong point of comparison for how landscape design can be understood. The designers organized the site circulation as a contiguous interlocking network and located points of rest at nodes of overlap in the grid. There is also an explicit hierarchy in the design. Consider for example the circulation system, where every 600 meters there is a vehicular bypass, at 300-meter

9 Miller, Wilhelm. *The Prairie Spirit in Landscape Gardening*. Amherst: University of Massachusetts Press, 2002, P5.

10 Miller, Wilhelm. *The Prairie Spirit in Landscape Gardening*. Amherst: University of Massachusetts Press, 2002, P17.

11 Tafuri, Manfredo and Gregotti, Vittorio. *Buildings and Projects*. New York: Rizzoli, 1982, P15.

中西部居民的需求，特点是通过保留本地色彩、重复使用在草原景观中最具代表性的天空与土地的水平线条，来保持典型的西部风情"。[9] 这显然是一种非常地域性的风格，其核心毫无疑问就是弗兰普顿所述的"平坦基质"。延森和西蒙兹花费了大量时间在他们的著作与设计作品中推行他们的主张——高差只是场地的一个特性，这个特性在环境力量（天气、日照、水文）与绿色植被面前都难以被操控。在这里，笔者并非试图将草原风格运用到浦江，而是想要强调以下观点：上海的气候、高水位以及长期受农业影响的土壤成分，都将共同决定本地的适宜植物群体，并将影响本地物种体验场所的方式。不管建筑风格如何，景观设计可以支持弗兰普顿所主张的批判性地域主义。"白纸状态"本身就是场地重要的历史文化元素。一个了解草原风格的景观设计会寻求保护并强化这种长而广的视野、运用本地植物、微妙地调整土地形态从而控制微气候——或者说得更形象一些："外来的人们还没有聪明到能看出在这种'平坦'草原风格中的轻微起伏，而这种起伏给予生活在这片土地上的人们无穷享受。"[10]

这并不是对格雷高蒂国际设计公司的浦江方案的批判，更多的是对其被描述和接受的方式的客观评价。事实上，该公司正是因其对于地域特殊性的强烈主张而著称，有评论称其"*像理解考古遗迹那样去理解设计场地*"[11]。如果景观设计如同高差、环境与植被一样被纳入城市设计的范围，并在建筑设计之前进行——就像格里高蒂的浦江新镇方案中的框架格网一样——那么快速城市化进程中的文化连续性、生活宜居性、环境可持续性将会大大提高。

生态既是一种区域基础设施，又是一种精确的设计操作

对格里高蒂国际设计公司的框架格网的描述，对于景观设计如何被理解是一个很好的比较。设计者将场地流线组织为一个互相咬合的连续系统，并将供行人休息的开放空间布置于道路交叉处，与格网重合。另外，设计方案中可以看出明确的层级。例如流线系统，每600米有一条机动车分流道，每300米有一条人车混行道

---

9 威尔海姆·米勒，《景观园林的草原精神》，阿姆赫斯特，马萨诸塞大学出版社，2002年，第5页

10 威尔海姆·米勒，《景观园林的草原精神》，阿姆赫斯特：马萨诸塞大学出版社，2002年，第17页

11 曼弗雷多·塔夫里，《维托里奥·格里高蒂：建筑与项目》，纽约：里佐利出版社，1982年，第15页

intervals are found roads, and at 150-meter intervals are found pedestrian pathways. Or, in terms of open space, we find intra-block courtyards giving way to inter-block plazas, and subsequently to the park which forms the central East-West spine of the project. The hierarchical organization is density dependent; at each successively larger scale, more citizens are planned to be moving though the space, and as such the scale of the circulation, the open space, and the architectural type responds in kind. Many years earlier Vittorio Gregotti wrote in *Casabella* that a primary aim for much of the firm's work was to identify "*a strategy of the discontinuous and of the circuit… based on diversification.*"[12] At the planning scale and in the initial phases of a project the search for this strategy should also be applied to landscape ecological systems. In every case the proposal need not express a formal and continuous open space system in order to assert itself as making sustainable improvements to the environment. Distinctions between a systemic and a discontinuous ecological system should recognize habitat ranges of projective future species. We should also recognize that landscape systems need not accept the same hierarchical organization as infrastructural or architectural systems. For example, there is advantage to infiltrating storm water locally, where it falls, instead of aggregating disparate catchment areas into a series of larger and larger ponds, or worse, hardened cisterns.

This way of thinking about the landscape, as an ecological system which operates in parallel to the urban structure, is extremely productive in light of two of the stronger critiques of Pujiang New Town. Firstly, while Gregotti Associati's plan proposed an intra-block pedestrian circulation system complete with accessible open space, in reality the public routes were rendered inaccessible, isolating each block and forcing pedestrian circulation to navigate parallel to the vehicular traffic and outward façade of the courtyard-type architecture.[13] Secondly, while it has been noted that the sale of units in Pujiang New Town was very successful, many purchases were completed as profit-driven investments; as a result only approximately 20 percent of the units have been occupied.[14] Urbanistically of course these unforeseen events are deleterious to the plan and to the success of the new town, but these failures also present an interesting opportunity for reconsidering the function of an urban ecosystem. For example, while pedestrian circulation may have been impinged by the privatization of the internal circulation of the blocks, neither environmental flows nor species movement patterns will likely be markedly affected by this modification of the plan. Indeed, one would project the opposite – that the internal courtyards will be far less heavily trafficked and as such could serve as oases in the dense urban fabric. Considering also that many of the units are at the present time unoccupied leads to a landscape type that should be low maintenance and need not necessarily be so formally expressive. There is an opportunity to firmly

12 *Casabella*, no. 421, P60.

13 Den Hartog, Harry. "Pilot Cities and Towns: Pujiang New Town (Minhang District)", in Harry den Hartog (ed) *Shanghai New Towns*. Rotterdam: 010 Publishers, 2010, P152.

14 Li, Xiangning. "Heterotopias: Themed Spaces in Shanghai and Los Angeles," in Harry den Hartog (ed) *Shanghai New Towns*. Rotterdam: 010 Publishers, 2010, P232.

路，每150米有一条人行道路；又例如开放空间，尺度最小的是街区内部庭院，其次是街区之间的公共广场，而尺度最大的是形成东西向中心轴线的中央公园。这种分级式的空间组织反映在建筑密度上，随着空间尺度增大，居民需要穿过的距离增加，因此交通流线、开放空间与建筑类型的尺度也随之增大。许多年前，维托里奥·格里高蒂曾在《美丽家园》（Casabella，意大利著名建筑杂志）上写到，格里高蒂国际设计公司的作品的主要目标是界定"*一种基于多样化的、将不连续空间与连续回路相结合的策略*"。[12] 在规划尺度上与在项目初期阶段，这种策略同样应该被运用到景观生态系统。在任何情况下，方案都不需要表现一种正式的、连续的开放空间系统，去证明自身对环境可持续性的提升。一个生态系统是否成体系，主要依靠未来目标物种栖息地范围的大小来判断。我们也应该认识到，景观系统不需要接受与基础设施或城市建筑系统相同的分级组织。例如，一种对生态系统有利的做法是使雨水在原地向地下渗透，而非将不同流域的雨水收集到一个更大的水池中，而硬化贮水设施则是更糟的选择。

上述的思维方式，将景观作为一种与城市结构平行的生态系统。根据对于浦江新镇的以下两条更有力的评论，这种思维方式是非常高效的。第一，虽然格里高蒂国际设计公司的方案规划了一个街区内部包含可达开放空间的人行流线系统，事实上公共穿过性道路是不可达的，并且这些道路将街区彼此隔离，迫使行人流线与机动车流线平行，从庭院建筑外围穿过。[13] 第二，虽然人们注意到浦江新镇的住宅销售情况十分成功，但大部分购买者只是基于投资需求，因此浦江新镇的入住率只有20%左右。[14] 当然，从城市的角度来看，这种不可预知的结果对于新镇方案的成功是有害的，但这些失败也揭示了一个有趣的机会，使我们重新思考城市生态系统的功能。举例而言，虽然人行流线可能被街区内部流线的私有化所侵害，但环境的流动与物种的迁徙规律都不会被这一方案改动所影响。事实上，人们可以提出相反的观点——即内部庭院的人流量会大量减少，因而成为密集城市空间中的绿洲。考虑到现在大部分单元还处于空置状态，我们可以得到一种景观类型，需要更低的维护成本，而不需要在形式上如此有表现力。我们可以看到一个契机，在城市区域达到其潜在的目标密度之前，稳固地建立一种有活力的生态

12 《美丽家园》，421期，第60页

13 哈里·邓·哈托格，《先锋城镇：浦江新镇（闵行区）》，摘自《上海新镇》，哈托格编著，鹿特丹：010出版社，2010年：第156页

14 李翔宁，《异托邦：上海与洛杉矶的主题空间》，摘自《上海新镇》，哈托格编著，鹿特丹：010出版社，2010年：第232页

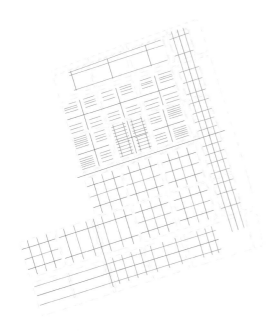

Building Fabric Composition
建筑布局肌理

— 空间虚轴 Void Axis
— 建筑实轴 Massing Axis

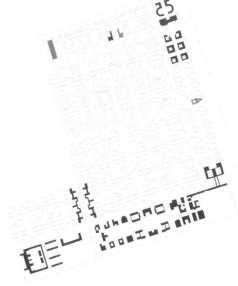

Urban Programs
城市建筑功能

居住 Residential
商业 Commericial
文教 Educational
水道 Canal

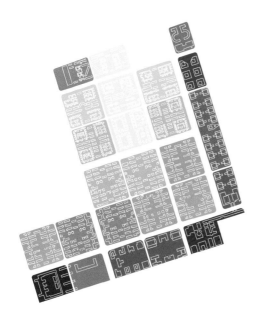

Floor Area Ratio
建筑容积率

3.2
0.2

Open Space Ratio
开放空间比率

93%
60%

establish a robust ecology before the urban area reaches its potential and projected density. An upfront investment in landscape design will pay dividend in the future. Over time a robust ecology will return ecosystem services, such as cleaner air, water, soil, and a healthier biodiversity. Additionally, in the near future, as demographics change, residents will look to leave the city center to occupy Pujiang New Town permanently. As homeowners they will come to appreciate the beauty of the established landscape, or alternatively, if selling a property they currently own, an established landscape will increase the value of the unit. This is ecology as a marketing strategy, timed to coincide with the eventual second wave of unit sales in Pujiang New Town.

The argument here calls for thinking about ecological urbanism at two scales. In the early design phases, landscape should be considered as an independent yet integrated system within the urban framework, where understanding and projecting environmental and species flows is most important. In later phases, as a detailed design exercise, landscape can contribute to enhancing an urban area's livability, cultural identity, and ultimately, its marketability.

Ecology can be simulated

One of the enduring critiques of Pujiang, and more generally of the government-planned One City, Nine Towns Development Plan of which it is a part, is the theming of these new development cores. In quick summary, we have the city of Songjiang as British, and the nine towns as follows: Anting - German; Buzhen - European/American; Fengcheng - Spanish; Fengjing - Canadian; Gaoqiao - Dutch; Luodian - Scandinavian; Zhoupu - European/American; Zhujiajiao - southern Chinese water town; and finally Pujiang - Italian. Not only were the designers from the themed country, but the construction in many cases constituted a copying of the most recognizable national architectural styles and often culminated in replicas of famous buildings and landmarks. This unrestrained replication is the subject of Bianca Bosker's *Original Copies: Architectural Mimicry in Contemporary China*, where she concludes from her work that *"the comprehensiveness of these copies has elicited criticism and derision on the part of Western and Chinese intellectuals alike, whose instinct is frequently to reject these themed communities as 'kitsch,' 'fake,' 'temporary,' or 'unimaginative and cliché.'* "[15] In her book, Bosker goes on to give a fascinating account of differing attitudes towards the production of copies in 'Western' and Chinese culture, building an argument that in Chinese ontology the real and the simulacrum need no differentiation – a very troubling concept for Western philosophy.[16]

15 Bosker, Bianca. *Original Copies: Architectural Mimicry in Contemporary China.* Honolulu: University of Hawai'i Press, 2013, P3.

16 Ibid. P25.

体系。这笔在景观设计层面的预先投资将在未来得到回报。随着时间推移，这一有活力的生态体系会将生态系统服务回馈给我们，例如洁净的空气、水、土壤，以及更健康的生物多样性。不仅如此，在短期的未来，随着人群变化，居民会离开城市中心，长期定居于浦江新镇。由于他们是业主，他们会欣赏这里的建成景观，或者如果未来他们卖掉持有的房产，这里的景观会提升房产的价值。这是生态作为一种市场营销策略的证明，在时间上正好符合浦江新镇房产销售的第二个高峰。

这一论点需要在两个尺度上思考生态城市主义。在设计初期，景观需要被看作一个在城市框架下独立而集成的系统，在这个阶段最重要的是理解与定位环境及物种的流动。在后期，随着设计的细化，景观可以提升城市区域的宜居性与文化特质，最终提高市场营销的程度。

生态可以被模拟

对于浦江新镇和"一城九镇"规划方案，长期以来一直存在一种针对每个新镇的主题化的批评声音。"一城九镇"中，"一城"指的是英国风格的松江新城，"九镇"则包括德国小镇安亭、欧美小镇陈家镇、西班牙小镇奉城、加拿大小镇枫泾、荷兰小镇高桥、北欧小镇罗店、欧美小镇周浦、南方水乡朱家角以及意大利小镇浦江。不仅这些小镇的设计者来源于上述主题的国家，而且这些小镇的设计手法也大量借鉴了该国家的建筑风格，甚至抄袭著名建筑地标。这种无限制的复制正是比安卡·博斯克《原创复制品：当代中国建筑模仿》一书的主题，她在书中总结到："*这些复制品之广泛和全面，在西方和中国学界都引发了批评与嘲笑，其中大部分都本能地否认这些主题城市，认为它们是'媚俗的''伪造的''临时的'或者'无法想象的、老套的'。*"[15] 在书中，针对中外对于这种复制品生产的不同态度，博斯克进一步给出了一个出色的描述，她认为在中国的本体论中，真实与虚假无需区分，这对于西方哲学而言是非常难以理解的。[16]

---

15 比安卡·博斯克，《原创复制品：当代中国建筑模仿》，火奴鲁鲁：夏威夷大学出版社，2013年，第3页

16 同上，第25页

In support of this argument Bosker cites Zong Bing's seminal essay "*Preface on Painting Mountains and Water,*" written on the subject of landscape painting and design where it is so noted that "*the Chinese did not impute significant differences between constructed landscapes in two-dimensional forms (i.e., paintings) and three-dimensional forms (i.e., imperial parks).*"[17] Without straying too far, it is well worth mentioning that this same reasoning underlies the early history of Western landscape architecture, where we would begin a follow up essay by citing Alexander Pope's proclamation that "*all gardening is landscape-painting; just like a landscape hung up.*"[18]

To the history of landscape painting Bosker also adds a phenomenological argument to collapse further the difference between a real landscape and a simulation. From our own experience we are certainly aware that if executed correctly a simulation – a painted view of a landscape – is able to elicit a mental and emotional reaction as deep as those produced by the real view.[19] And finally she discusses the Chinese history of collecting and importing plants to private hunting gardens from as early as the third century BCE.[20] Again it would be interesting to re-read the history of plant collection and transportation in China against later projects that would follow in the Western tradition, culminating in Joseph Paxton's Crystal Palace.

While Bosker uses Chinese attitudes toward landscape reproduction as an analogy to make an argument about urban design in China and the West, there is a strong case to be made for more scholarly research which focuses specifically on landscape simulacra. Not only could comparisons be made between landscape painting, phenomenology, and species collection, but the uniquely Chinese history of landscape copying would certainly add a new dimension to the discussion of environmental restoration. In the West, and especially in conservation science, there is a pervasive attitude that human modification of an environment is categorically different than natural modification; that we cannot author landscapes in the same way nature can. This thinking leads to a paradox as so described by Christine Biermann and Becky Mansfield: "*In short, conservation science aims to control life – all the while drawing the line between biological threats and advantages – in order for life to proliferate freely.*"[21] Of course this paradox only exists if we recognize a difference between human control of life and the control of life exerted daily by environmental forces. If human designed landscapes are accepted without prejudice as equally authentic to those arrived at by nature, then restoration

[17] Ibid. P26.

[18] Hunt, Dixon. *Gardens and the Picturesque: studies in the history of landscape architecture.* Cambridge, Mass.: MIT Press, 1992, P106.

[19] Bosker, Bianca. *Original Copies: Architectural Mimicry in Contemporary China.* Honolulu: University of Hawai'i Press, 2013, P27.

[20] Ibid. P30.

[21] Biermann, Christine and Mansfield, Becky. "Biodiversity, Purity, and Death: Conservation Biology as Biopolitics", *Environment and Planning D: Society and Space* 32 (2014): 264.

为了支持这一论点，博斯克引用了宗炳的有创造力的文章《画山水序》，后者以山水画与设计为主题，提到"*中国人并不想明确地区分二维图像（山水画）和三维空间（皇家园林）*"。[17] 本文并不想过多地涉及其他论题，但确实值得一提的是，早期西方景观建筑学也是基于相同的原因发展起来的，我们可以引用亚历山大·波普的宣言："*所有的园林都是景观画，就像把景观挂起来那样。*"[18]

对于景观绘画，博斯克还提出了一个现象学的观点，从而使真实与模拟景观之间的区别进一步瓦解。从我们自身的经验而言，我们确定地知道，一个恰当的模拟过程——比如一张关于景观的绘画——可以像亲眼得见一样引发深层的精神与情绪反应。[19] 最终，博斯克讨论了中国从公元前三世纪开始向私家狩猎园林中收集与引进植物的历史。[20] 再一次地，重新阅读这一部分历史并将其与"一城九镇"项目案例对比将会变得非常有意思，因为后者其实是西方传统的产物，就像将植物放进约瑟夫·帕克斯顿的水晶宫里培养。

虽然博斯克将中国对待景观复制的态度作为一种类比，来分析中国与西方的城市设计，但我们需要重点关注的是关于景观复制品更学术性的研究。我们不仅可以在山水画、现象学、物种收集之间做对比，还可以从中国独特的景观复制史中得到对环境修复的新思考。在西方，尤其是在保护科学领域，一种普遍的观点是人类对环境的改造与自然对环境的影响在本质上是不同的，人类不能像自然一样去改造景观。这一观点将我们引向一种矛盾，克里斯丁·毕尔曼和贝基·曼斯菲尔德这样描述这种矛盾："简言之，*保护科学的目标是控制生命——持续地在生物的威胁与优势之间划分界限——从而保证生命自由地繁衍。*"[21] 当然，只有当我们认识到人类对生命的控制不同于环境力量对生命的绝对掌握时，这种矛盾才会存在。如果人工设计的景观被认为像自然生成的景观那样具有同等的原真性，并被不带偏见地接受，那么生态修复工作就可以更加自由地进行，并产生与任何自然演化轨迹都迥然不同的崭新的未来。显然，我们处于人类世（Anthropocene），长期以来西方公认的保护概念将被削弱。在我们接受人类作为气候与环境变化主

---

17 同上，第26页

18 迪克松·亨特，《景观建筑学历史中的园林与如画美学研究》，坎布里奇，马萨诸塞州：麻省理工学院出版社，1992年，第106页

19 比安卡·博斯克，《原创复制品：当代中国建筑模仿》，火奴鲁鲁：夏威夷大学出版社，2013年，第27页

20 同上，第30页

21 克里斯丁·毕尔曼，贝基·曼斯菲尔德，《生物多样性，纯粹性，死亡：作为生物政治的保护科学》，摘自《环境与规划D：社会与空间》，2014年，32期，第264页

efforts would be provided with much more freedom to intervene and project new futures distinct from those on any evolutionary trajectory. As we begin to recognize our era as firmly within the Anthropocene, longstanding Western notions of conservation will carry less relevance. In accepting our position as dominant drivers of climate and environmental change, the authenticity of non-human forces will be harder to identify. If America is, as Umberto Eco describes it, *"a country obsessed with realism, where, if a reconstruction is to be credible, it must be absolutely iconic, a perfect likeness, a 'real' copy of the reality being represented,"* then China's alternative attitude to the copy may serve useful in understanding and accepting novel human designed landscapes.[22]

Conclusion

What we begin to argue for is the ascendency of landscape as a critically important design component of urban planning. Repeated again from the introduction, Kenneth Frampton described Gregotti's position on design as an "act of culture in the void of nature." Another project often noted as a paramount part of the lineage of city-making leading up to Pujiang and Shanghai's One City, Nine Towns Development Plan is Ebenezer Howard's Garden City. Gregotti's thinking is shared with Howard's, where a rational grid framework gives order and form to an urban development in an otherwise undifferentiated space of nature. However, it is becoming increasingly problematic to view nature as a surrounding void. We are increasingly aware that this void is being unintentionally designed by environmental forces from our economic and industrial activities. At the same time, the physical extent of any void space is rapidly decreasing, making it far less relevant to the lives of those within urban centers. Finally, the void is increasingly recognized to have an independent culture, one which continues to shape how it is that we mark the ground around it. If *"the city is the one environment created exclusively for human use"*, this way of thinking, as summarized by Yi-Fu Tuan, will have to adapt.[23] The city should also be designed for nature; to ignore this opportunity is ecologically and economically unsound.

22 Eco, Umberto. *Travels in Hyperreality.* New York: Harvest Book, 1986, P4.

23 Tuan, Yi-Fu. "Place: An Experiential Perspective," *Geographical Review* 65, no. 2 (1975), P157.

要推手这一角色的同时，非人工力量的原真性将更难界定。如果就像乌姆贝托·艾科描述的那样，"*美国是一个沉迷于现实主义的国家，这种现实主义表现为，如果要重建，那么一定是标志性的、一种完美的肖像，一个现实的'真实'复制品*"，那么中国采用的对待复制品的另一种态度也许对于理解与接受人工设计景观是有用的。[22]

结论

本文开篇提出这一观点，即景观应被作为一种城市规划中尤其重要的设计组成部分的优势。这里再重复一次，肯尼斯·弗兰普顿将格里高蒂对设计的态度描述为"*在自然的空白中的文化行为*"。当人们讨论上海"一城九镇"规划方案时，常常提起埃比尼泽·霍华德的城市规划经典之作——花园城市。格里高蒂的思考与霍华德不无相似之处，即通过理性的格网框架带来秩序，在缺少限制的自然空间中塑造城市发展。然而，"将自然看作一种包围城市的空白区域"这一观点正越来越多地表现出问题。我们越来越清楚地了解，这种空白是在我们的经济与工业活动中由环境力量无意中制造的。与此同时，这些空白领域的范围正在快速减少，使之与城市中心人群的生活越来越不相关。最终，人们逐渐认识到这种空白领域发展出一种独立的文化，随着人类城市逐渐扩张，这种文化却一直发展着。"城市是为了人类的需求制造出来的具有排外性的环境"，段义孚总结的这种普遍的思维方式亟待改变。[23] 城市同样应该为自然而设计，忽视这一机会在生态和经济层面都是不完善的。

22 乌姆贝托·艾科，《超现实旅行》，纽约：哈维斯特书店，1986年，第4页

23 段义孚，《场所：一种实验性视角》，《地理研究》，1975年，65期（第2本），第157页

Pujiang New Town is located in the south of central Shanghai, proposed in the "One City, Nine Towns" initiative with the purpose of catalyzing the integrated development of urban and suburban areas. In 2001, Tianhua worked in collaboration with Gregotti International, who won the design competition; they took from Italian Rationalist urban planning theories and developed a conceptual proposal following the urban fabrics and spatial logics of Italian towns. The plan takes the canals as its arteries and distributes commercial, office, municipal, cultural, and educational programs along the canals. Each block is subdivided by green space, and buildings are all connected by internal pedestrian walkways. The new town has become the bridge between Puxing Highway and the Huangpu River. So far, Pujiang New Town has been partially completed as a new suburban habitat, but the urban infrastructure has not been established and the core area has not functioned well enough to draw as many residents as expected. It remains a radical issue to improve the conditions and solve the paradox of Pujiang.

浦江新镇位于上海市区南部，背景为"一城九镇"政策，旨在带动上海城郊与城市中心的一体化发展。2001年，天华与赢得竞标的格雷高蒂事务所合作，采用意大利理性主义城市规划经验，借鉴意大利小镇的城市肌理与城市空间逻辑对概念方案进行调整深化。浦江新镇围绕中心河道展开，沿河布置了商业、办公、市政中心、文化中心、大学等城市功能空间，每个街区单元都被绿化带分割成小块，再由步行街道连接每小块的建筑物，并逐渐延展成城市街区，整个新镇成为贯穿浦星公路至黄浦江的纽带。目前，浦江新镇已形成规划中的新型城镇雏形，但城市基础设施尚未建立，核心区域功能缺失，导致区域入住率较低。如何改善与解决浦江新镇困境将是一个长期的城市难题。

Yao Xiao
with Dingliang Yang

Building the Form of a Territory
A Case Study on Pujiang New Town

Background

In 2001, the Shanghai government announced a policy to build "One City and Nine Towns" in its suburban areas, with the intention of introducing European townscapes by involving architects and planners from European countries. Among the ten sites, Pujiang New Town has neither the significant advantage of industry, as in places like Anting New Town, nor the historical resources of sites like Fengjing or Zhujiajiao; however, it does benefit from its proximity to the city center of Shanghai and convenient public transportation.

Pujiang New Town actively implemented the government policy, using international design competitions to select the most suitable master plan. The organizer invited three design firms associated with practice in Italy to submit master plans for the entire 15 km² area and also promised that the winning firms would be simultaneously awarded the design of buildings within Pujiang New Town. The "highly risky, but at the same time madly fascinating" scheme from Gregotti Associati International eventually won the competition over the "collage of historic Italian building clusters" from Scacchetti (Fig.1) and "water linked American suburbs" from SWA (Fig.2). Gregotti's proposal was selected for its merits of rational planning by building close connections between the site and the city, creating compact urban form and well-designed building types, as well as the promotion of public space and the encouragement of public life (Fig.3). It was called "risky" because the proposal presents an Italian town through a reinterpretation of the traditional model that reflects the spatial spirit of 'Italian Town,' rather than a straightforward carbon copy of the traditional or vernacular townscapes of Italy, as the winning projects of other New Towns had done. From the perspectives of professionals, this 'risk' is worth trying and a very good move.

Though most of the projects within the One City, Nine Towns initiative were described and criticized as "theme park-like developments" by designers and the media, after the publication of the master plan from Gregotti, it was widely accepted as *"a rational design, instead of a theme park-like 'foreign architectural exhibition.' "*[1] (Fig.4) After winning the competition, Gregotti Associati International paired with Tianhua as the local design institute to continue the detailed urban design, building design and infrastructure-planning on the 2.6 km² plot. These two actors contributed to the development of the proposal by taking advantage of their own deep understandings of both Italian culture and Chinese domestic conditions. According to the 2003 description from Augusto Cagnardi, the design director of Gregotti Associati, *"the foreign and Chinese parties involved in the development all feel good about keeping their own viewpoints and compromising in some common interests."*[2]

[1] Xue, Charlie QL, and Minghao Zhou. "Importation and adaptation: building 'one city and nine towns' in Shanghai: a case study of Vittorio Gregotti's plan of Pujiang Town." *Urban Design International* 12, no. 1 (2007): 21-40.

[2] ibid.

肖瑶
及杨丁亮

建设区域形态
浦江新镇案例分析

背景

2001年，上海政府颁布了在上海郊区建设"一城九镇"的政策，希望（新城新镇）能塑造出欧洲城镇的面貌，计划邀请来自欧洲各个国家的设计师和规划师来参与规划设计以他们自己国家命名的城镇。在十个城镇项目中，浦江新镇虽然没有像安亭新镇那样的产业优势，也没有像枫泾和朱家角那样的历史资源，但是靠近上海城市中心的地理位置和便利的公共交通设施给它带来了自身的优势。

浦江新镇积极落实了政府的政策：通过国际竞赛招标来选出最合适的规划方案。竞赛的组织者邀请了三家在意大利有着广泛实践的公司来为15平方公里的整个区域提出构想，并承诺竞赛胜出的单位也会同时获取设计浦江新镇内部建筑的设计权。来自格雷高蒂国际建筑设计公司"具有高度挑战性、但又同时极具吸引力"的方案胜过了来自斯卡切蒂的"意大利历史街区拼贴型设计方案"(图1)和SWA的"水系连接的美国郊区型设计方案"(图2)。格雷高蒂的提案能获胜，很大程度上是因为它通过理性的规划，把设计基地和上海规划中的城市环境建立了密切的连接，创造出一种低层高密度的城市形态，设计了优秀的建筑形制，并且通过提倡对公共空间的设计来鼓励居民对于公共空间的使用。(图3)说这个方案有挑战性是因为它没有像之前几个新镇规划设计竞赛获胜的方案那样进行粗暴的复制，它没有直接模仿意大利建筑和街区的面貌，而是试图去重新理解传统的模式并通过塑造意大利城镇的空间感受来设计和反映意大利城镇的精髓。但是从设计从业者的角度来看，这是一种值得的冒险，而且是一种非常好的方式。

尽管"一城九镇"计划中的大部分新镇项目被媒体和设计师们描述或批判其实为一种"主题公园型的开发"，但是在格雷高蒂的规划设计方案公开之后，大家普遍认可和接受这一方案，认为它是"*一个理性的设计，并非是像外国建筑展览形式的主题公园设计*"。[1] （图4） 格雷高蒂国际建筑设计公司继续与本土设计院天华合作开展2.6平方公里的指定区域进行深入的城市设计、建筑设计和基础设计规划。双方充分利用各自对于意大利设计文化和中国的具体情况的深入了解，对方案的推进做出了贡献。根据格雷高蒂国际建筑设计公司的设计总监卡尼阿蒂的描述，"*国外和国内设计方都支持双方在一些主要的方面达成一致的共识，（而对于其他方面）各自保持自己的观点*"。[2]

1 薛求理，周鸣浩，《引入与适应：上海"一城九镇"的实践：格雷高蒂的浦江新城案例分析》，《国际城市设计》2007年第12期：21-40.页

2 同上

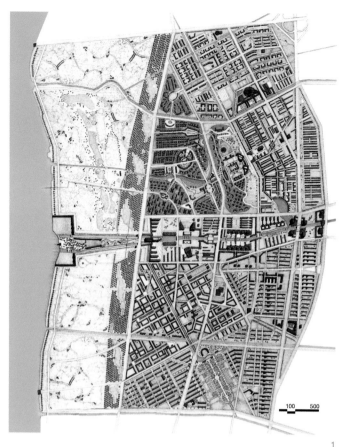 

Fig.1 Masterplan for Pujiang New Town, SWA Group, 2004
Fig.2 Masterplan for Pujiang New Town, Scacchetti Associati, 2004

图1　浦江新镇规划设计详细图，斯卡切蒂设计事务所，2004
图2　浦江新镇规划设计图，SWA集团，2004

Toward the form of the territory

Before Pujiang, Gregotti had completed several large-scale projects, including the Bicocca district in Milan. However, the scale of Pujiang New Town is beyond the urban scale in a traditional sense; it is closer to a territorial scale creating a new city. In the design of this project, we can clearly see the impact of Vittorio Gregotti's theoretical thinking on the form of large-scale urban settlements and landscape, which he later defined as "the form of territory." As a pioneer of architectural thinking on territory, Gregotti aims to define formal methodologies and approaches adaptable to different scales. *"We must first determine to which extent the issues of territorial planning, considered from the point of view of the specific codification problems that they imply, have generally shifted the problematic of architectural space by elevating it to the level of geographic space. Next, we will have to find the means of intervention corresponding with different scales."*[3] According to Gregotti, in order to find the means, or the 'new design instrument' as we can call it here, architects need to absorb knowledge from other disciplines, since traditional architectural means are not well-adapted to large-scale projects and interventions. Further, Gregotti defines a space he calls the "field," which is somewhat independent of scale, and can be constituted as an element of a wider ensemble or as a macro-structure that includes a series of fields or ensembles. Here we understand the theory; Gregotti tries to find a form of scalability that can configure space, which he calls both "field" and "ensemble," a territory but at the same time the unit and prototype of the territory. This idea breaks from the confines of the classic idea that the form of architecture is always deeply related with its material and functionality. Importantly, Gregotti finds a unique formal relationship between geometry and geography.

In Pujiang New Town, the planners adopted the grid as a basic form to structure the entire territory. Using the grid as a device perfectly coincides with Gregotti's theory and

3 Gregotti, Vittorio. "The Form of the Territory."
OASE Journal of Architecture 80 (2010).

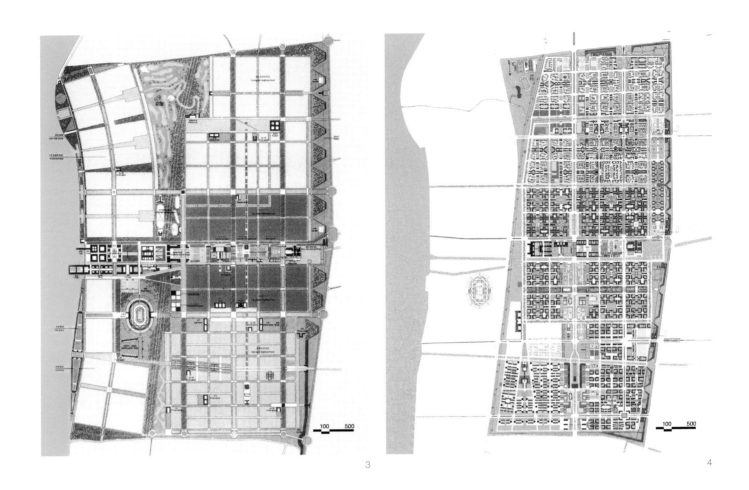

Fig.3 Conceptual Masterplan for Pujiang New Town, Gregotti Associati International, 2004
Fig.4 Detailed Masterplan for Pujiang New Town, Gregotti Associati International, 2007

图3 浦江新镇规划设计概念图，格雷高蒂国际建筑设计，2004
图4 浦江新镇规划设计详细图，格雷高蒂国际建筑设计，2007

走向区域的形态

在浦江这个项目之前，格雷高蒂已经完成了许多大尺度项目，其中包括米兰的比可卡这样的著名项目。然而，浦江新镇所要求的设计尺度依然远超了传统理解中的城市尺度，更接近塑造城市般的一种区域尺度。在这个项目的设计中，我们可以十分清晰地看到格雷高蒂对于大尺度城市设计和景观的理论思考所产生的影响。格雷高蒂之后将自己的（大尺度城市设计）思考归结为"区域形态"。作为建筑师人群中思考"区域形态"的先锋人物，格雷高蒂想要去定义一种适用于各种不同尺度的形式设计方法 "从具体的定义问题的角度考虑，我们首先需要确定到何种程度，建筑空间的问题可以转化为区域规划，并且上升到一个地理空间的尺度。然后，我们就必须去找到一种介入的方式用以回应不同尺度（的需求）"。[3] 根据他自己的说法，由于传统的设计方式不适用于大尺度的项目，建筑师必须吸收其他学科的知识，这样才能找到和发现新方法，这里我们称之为"新的设计工具"。之后格雷高蒂定义了一种新的没有尺度概念的空间，他称之为"场域"。这个所谓的"场域"可以组成一个更广大集聚空间的单元，也可以是一个包含了一系列场域或者集聚空间的巨大空间体系。在这里，我们可以把这个理论理解为格雷高蒂试图去寻找一种可以塑形空间又具有延展性的形式，去适应他所说的"场域"和"集聚空间"，一种既是区域又是区域单元的空间。在这里，（这种理论）突破了之前关于建筑的经典理论，从而使建筑形态不再从根本上与建筑材料和功能绑定。重要的是，格雷高蒂发现了几何与地理之间的一种独特的形态关系。

[3] 维托里奥·格雷高蒂，《区域的形态》，《OASE建筑杂志》总第80期，2010年

intentions of making form fit responses to different scales. As Rosalind Krauss declares, *"Grid states the absolute autonomy ... Grid is about being, mind, or spirit, is a staircase to the universal ... because of its bivalent structure; it is both centrifugal and centripetal existence."*[4] In this sense, the grid as a form of geometry can extend in all directions to infinite scale, and at the same time can be scaled down continuously. It is so powerful that it becomes the form of territory, the form of the block, and even the form of architecture in the design of Pujiang New Town.

Geometry as the form of Territory

"Designing a territory is always treated as the colonization of nature by man."[5] Gregotti describes the process of design as a territorial project on a *tabula rasa*, a way of conquering nature, which explains why he chooses to use geometry as the form of territory to illustrate the extreme contrast between the man-made and the natural. Geometry as a design device was used in the neo-rationalist Italian architecture movement of the 1960s. The design of Pujiang New Town employs "the ideas of Tendenza", using the grid as the form to shape the entire 12 km^2 territory. The scalability of the grid allows it to act as a prototypical geometry to organize the space at the planning scale but also at the block scale (Fig.5). The form of geometry means strict order and clear division. It is fair to say the form of Pujiang New Town is a three-dimensional gridded city. At the spatial-structural level, the content of open space and infrastructure are major actors in the grid system, a duality shaped by the canal system and greenway system. They conceive the spatial hierarchy by laying out two perpendicular central axes framed by wide canal- or tree-lined axial avenues, along with a series of roads and greenways, forming a land-mesh that divides the territory into 14 x 5 x 300-meter blocks. *"A framework grid was prepared in combination with strict urban rules, which according to Italian tradition guarantees architecture variety."*[6] The horizontal axis accommodates public parks and landmark buildings, including a shopping mall, governmental buildings, a university, and the promotion center, all designed in the form of the grid. Perpendicular to the horizontal axis is the vertical, natural axis, consisting of canals and greenways. For the general block structure, each 300 x 300 meter square is surrounded by roads, forming a different community with its own identity (Fig.6). Inside each block are four access routes, all consisting of pedestrian and bicycle paths.

Rigid geometry as the form of Pujiang New Town is reflected not only in the strong figurative objectives of the design, but also in the strict 'geometrical' control of the practical

---

4 Krauss, Rosalind. "Grids." *October* 9 (1979): 51-64.

5 Gregotti, Vittorio. "The Form of the Territory." *OASE Journal of Architecture* 80 (2010).

6 Den Hartog, Harry, ed. *Shanghai new towns: searching for community and identity in a sprawling metropolis.* Rotterdam: 010 Publishers, 2010.

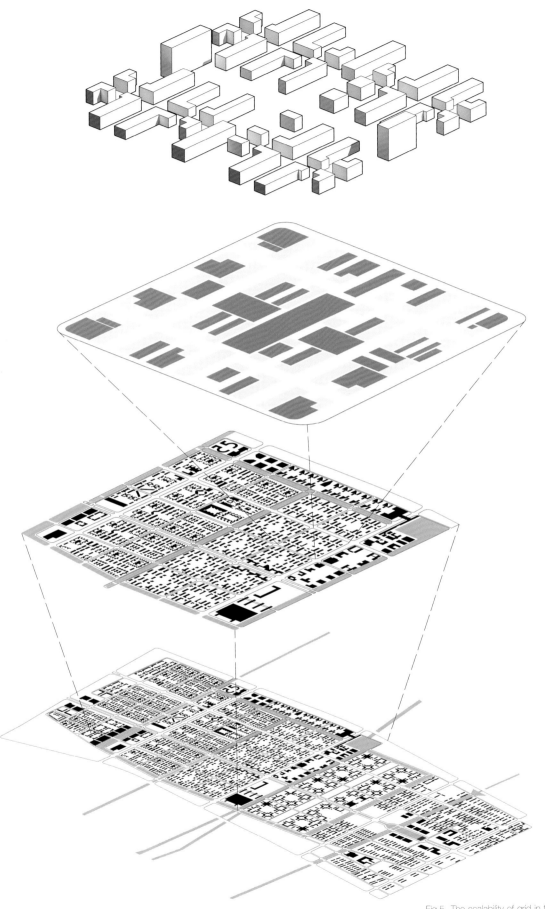

Fig.5 The scalability of grid in the project of Pujiang New Town
图5 浦江新镇规划设计中网格的延展性

在浦江新镇的规划中，采用了网格作为该区域的基本形态骨架。采用网格作为手段的举动非常契合格雷高蒂关于设计形态以适合不同的尺度空间的理论和意愿。正如罗莎琳德·克劳斯所描述的，"*网格表达了一种绝对自主……网格是关于存在、想法和精神的，是通向宇宙的阶梯……因为它具有双向的结构；是一种既是向心，又是离心的存在*"。[4] 在这种情况下，网格作为一种几何形态可以向各个方向无穷延展，也可以不断向内缩小尺度。网格非常有力，从而使得其在浦江新镇的设计中既成为区域的形态，又成为街区的形态，甚至成为建筑的形态。

几何形式作为区域形态

"*设计一个区域总会被认为是人类在占领和殖民自然。*"[5] 格雷高蒂这样形容，他把在原生态的田地上勾画一个区域性项目的过程比作是在征服自然。这也很好理解他为什么选择集合形式作为区域的形态：因为他想让人为塑造的和自然存在的（东西）形成强烈的对比。几何形式在六十年代左右的意大利的新理性主义的建筑运动中也被用作（主要的）设计的手段，浦江新镇的设计运用了当时的"坦登扎"理念，把网格作为主要的形态来覆盖整个12平方公里的区域。网格自身的形态可伸缩性也使得它可以作为基本的几何单元从规划的大尺度和街区尺度来组织空间。(图5)几何形态意味着严格的秩序和清晰的分区。可以说浦江新镇的形态就是一个三维网格城市。在空间结构层面，主要是开放空间和基础设施在网格系统中起作用。道路系统和河道系统形成了二元网格体系。他们通过宽河道和林荫道划定出两条相互垂直的中心轴线，继而确立了区域空间层级；再加上一系列道路和绿轴，形成了一个土地网格把整个区域分成了14乘5个300米大小的街区。

"*根据意大利传统设定的一个空间网格框架，再赋予其严格的城市空间规则，就保证了其中建筑的多样性。*"[6] 横向的中轴上主要容纳了公共公园和包括购物中心、政府大楼、大学和推广中心等在内的地标性建筑，而且它们都被设计成了网格的形式。与横轴垂直的是由河道和绿轴组成的垂直中轴。对于普通的街区，一个300米乘以300米的正方形街区形成一个社区，各有各的特点。(图6)它们由道路包围，其内部有4条人行道路伴以自行车道。

---

4 罗莎琳德·克劳斯. 《网格》，《十月》杂志 1979年第9期：51-64页.

5 维托里奥·格雷高蒂，《区域的形态》，OASE 建筑杂志总第80期，2010年

6 哈里·邓·哈托格（编），《上海新城：追寻蔓延都市里的社区和身份》，鹿特丹：010出版社，2010年.

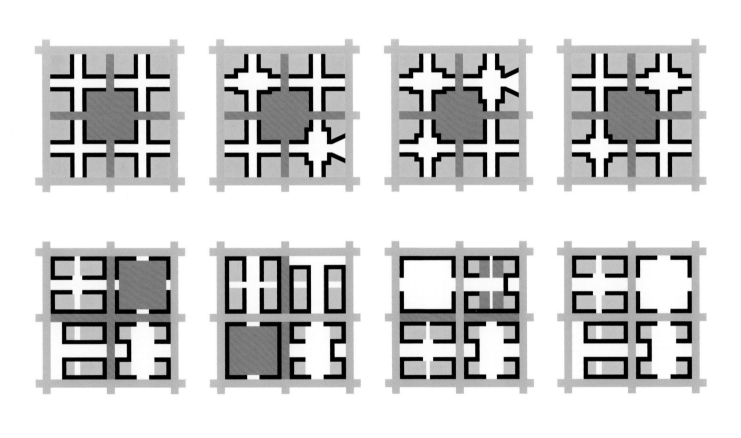

住宅 Residences
步行街 Pedestrian Road
私家花园 Private Gardens
绿色步行街 Green Pedestrian Road

Fig.6 The morphological typology of blocks
图6 街区的肌理形式类型

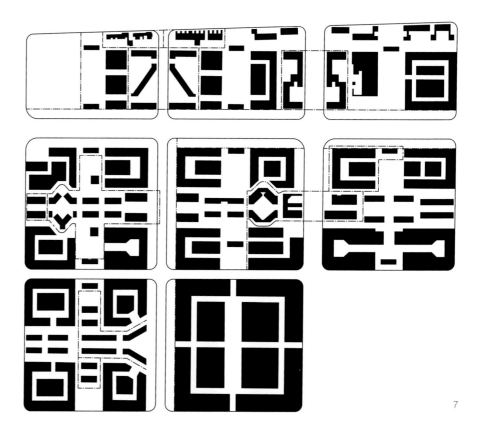

Fig.7 Block design for 49 hectare of flagship district by Gregotti Associati International and Tianhua

图7 格雷高地国际建筑设计和天华建筑设计领导的的49公顷示范区街区设计

management of different prototypical blocks. Each block as a geometry is designated to host different groups of people. The geometry turns out to be a fragmented landscape, with each square as a gated community socially distinct from other squares and with segregated classes. This actually conflicts with the initial master plan's intention to mix functions and housing types with different prices and income levels, which is essential to European town planning. However, to an extent, the geometry itself represents a certain order, both spatially and socially, which concentrates on the territorial geometry, echoing the idea of Versailles from the 17th century. In China, however, this kind of segregation in form and class is aggregated under the ideology of stereotype, so that people somehow reject living in neighborhoods with a social mix.[7] The negative effect of the enclosure of the blocks also impacts the continuity of urban space and the itinerary, which we argue harms the accessibility and connectivity of the territory as a whole.

Architecture as the form of Territory

In theory, Gregotti thinks that the form of a territory could be interchangeable with the form of architecture to the proper extent. Hence, after the gridded plan, the design of Pujiang New Town quickly zooms in to the level of building dimension and form. Designing the form of the ensembles, helps to better shape the whole plan. Here if we understand the form of architecture then we can somehow comprehend the form of the territory, since they are so deeply linked in Gregotti's proposal and thinking.

An area of 2.6 km^2 was assigned as the exemplar for the entire New Town, and the project stands for more diversity within residential blocks, which are designed with four main dwelling types: Courtyard Houses based on Domus Pompeiana, Waterfront Townhouses, Riverside Houses, and Garden Condominiums. Further, 49 hectares of 8 blocks are designed in detail, with buildings and open space, to illustrate the texture and mixture that has been set as the flagship district when the master plan touches the ground (Fig.7). The concept of mixture manifests in the cross-scale design, and is most visible in the mixed-use buildings. Within the vision of a mixed community, Gregotti pays special attention to mixed-use buildings, such as the Promotion Center in Pujiang;

[7] Fokdal, Josefine. Bridging Urbanities: Reflections on Urban Design in Shanghai and Berlin. *LIT* Vol. 17, Verlag Münster, 2011.

浦江新镇高度几何的区域形态，不仅反映在强烈的（几何）图形化倾向的设计中，也表现在不同的单元正方形街区运转中严格的"几何"控制。每一个街区如同一个图形一样被指定安置不同的人群。几何形态变成了一种分割开的景观，其中每个方框就是一个区分社会层次的封闭式小区。这实际上是与最初的规划理念相违背的。最初的设想是遵循欧洲城镇规划的本质，把不同功能的建筑、不同形式和不同价位的住宅混合在一起，从而把不同收入、社会层次的人群在生活中融合并区分到各个街区之中。然而，从某种程度上说，几何形式自身本来就带有一定的空间和社会等级的区分，而这种特性就更进一步地集中体现在区域的几何形态之中，就比如17世纪的凡尔赛。而在中国，这种形式上和社会层级上的分隔在老套的意识形态下被放大了，人们似乎拒绝在自己居住的社区中有混合的社会层级。[7] 这样的封闭小区带来的负面影响也表现在打断了城市空间和道路的连续性上，（对于这样的影响）我们需要强调它伤害了区域空间作为一个整体的可达性和联系。

建筑形式作为区域形态的一种呈现

在理论研究中，格雷高蒂认为区域的形态和建筑的形态在合适的范畴内应当是可以互换的。因此，在有了网格规划之后，方案的设计迅速深入到了建筑尺度的空间和形态设计。通过设计关键节点（建筑），来帮助更好地提升整体规划的水平。从这个思路出发，因为周边建筑和区域的形态与格雷高蒂的方案和思考紧密相关，所以我们可以通过了解设计中建筑的形态来了解区域的形态。

2.6平方公里的区域被设定成这个新镇项目的示范区，它的住宅范围的设计主要由四种建筑类型构成：庭院式住宅，水滨联排，河畔洋房以及公寓。继续深化，8个街区组成的49公顷的范围进行了建筑和开放空间的深入设计，作为示范区来展示如何将规划方案中的混合社区构想落地。(图7) 混合的概念跨越多个尺度贯穿于整个项目设计之中，集中体现于功能建筑。在混合的街区构想中，格雷高蒂特别提到了混合功能的特别建筑，比如浦江新镇的推广中心，他将其称之为"宫殿"

7 约瑟芬·弗克达尔.《连接城市：对上海和柏林城市设计的反思》，《LIT》第17期， 明斯特出版社，2011年

Fig.8 Semi-aerial photo of Pujiang Promotion Center
Fig.9 Site plan of Pujiang Promotion Center
Fig.10 Courtview of Pujiang Promotion Center

图8 浦江新镇推广中心半鸟瞰照片
图9 浦江新镇推广中心总平面
图8 浦江新镇推广中心庭院照片

he calls these buildings 'palazzos' and treats them as the epitome of the form of the territory. *"We reckon, in accordance with the tradition of the Italian historic cities, that such levels have to be mixed and that the different needs are taken into account within the context of the specific architecture: the 'palazzo' is an element of expression of the urban texture in the same way as the various house and apartment typologies."* [8]

The Promotion Center, located in the northwestern corner of the 2.6 km² exemplar district, follows the general order of the grid, aligning its built forms with it. Though the plan of the building looks regular, through the mixture of several tectonics and types of volume the building in three dimensions expresses an exquisite spatial richness and reflects the taste of the ancient Roman agora in modern architectural language (Fig. 8-10). *"The project is composed of the clubhouse, retail, office, exhibition hall, and a landmark bell tower. The external court is the core with the richest expression. It is divided into four huge quadrangles by a stark, thick red wall, elevated above people's heights. The concrete wall is clad with grey slates outside and painted a red color inside. Within the court, people's vision is naturally led to the sky. The blue sky, white cloud and red wall orchestrate an extremely pure picture."* [9] This modern form, based on the fantastic relationship between indoor and outdoor space, recalls the feeling of cityscape, while the wall, the pilotis, and the open space are reminiscent of Aldo Rossi's spatial treatment and his classical concept of architecture and the city, where architecture as a form, reflects the idea of the city.

Unlike for Gregotti, for us the most typical form of architecture that can represent the form of the territory of Pujiang is the housing. Normally when we look at the bar-shaped residence or the point tower, we think they must be a form of a city or a certain district because the image of Chinese cities filled and formed by ubiquitous north-south facing buildings is so entrenched in people's minds. In Pujiang New Town, most of the buildings are residential, and the proposal shapes the typology of residential blocks, superimposing them on the site into the urban structure, which is another reason why we think dwellings are the most representative form. Interestingly, the designers have used a consistent, systematic structure for all the blocks, but invented a more flexible type combining the villa, rowhouse, and apartment for residential blocks (Fig.11-12). According to the specific program requirements, buildings can modify accordingly; for instance, a building can be set containing more features of a villa and then be sold as a garden villa in the market. This kind of design gives people the impression of a residentially-dominated mixed and textured neighborhood, and also of a mixed and modern Sino-Italian Town, which is actually the real territorial form of Pujiang New Town.

8 Morpurgo, Guido, and Gregotti Associati. *Gregotti & Associates : The Architecture of Urban Landscape*. New York : Rizzoli, 2014.

9 Xue, Charlie QL, and Minghao Zhou. "Importation and adaptation: building 'one city and nine towns' in Shanghai: a case study of Vittorio Gregotti's plan of Pujiang Town." *Urban Design International* 12, no. 1 (2007): 21-40.

Fig.11-12 Garden Residence of Pujiang New Town
图11-12 浦江新镇花园式住宅

并认为它们是区域形态的缩影。"我们认为，按照意大利历史城市的传统，在特殊建筑的语境中，必须要有这种层次的混合，不同的需求都要被考虑在内。'宫殿'也是一个和各种房屋和住宅类型一样表现城市肌理的一个要素。"[8]

展销中心位于2.6平方公里的示范区的西北角，它的形态遵循着大的网格秩序，并沿着网格的边界塑形。尽管建筑的平面看起来不是很特别，但是通过各种建筑结构语言和形体类型的混合使用，它呈现出了一种特殊的空间丰富性，并用现代建筑的语汇反映出罗马阿格拉的一种古典的品位。(图8-10)"这个项目由俱乐部、零售、办公、展览和地标性的钟楼组成。有着丰富表现形式的外部庭院是整个建筑的核心，它由四条一面是红色，一面是灰色的悬在空中的梁限定成了四个方形的区域。在中国庭院，人们的视线被很自然地导向空中，蓝天、白云、红墙相互掩映，形成了一幅如画般纯净的场景。"[9] 这样通过现代的形式表现出杰出的室内和室外的关系，很容易让人联想到城市的风景（也是室内和室外的对比、融合）。而建筑中的墙面、柱网和开放空间唤起了人们对于阿尔多·罗西的空间处理方式和他经典的"城市-建筑"理念的记忆：建筑是如何作为一种形式来反映城市理念的。

与格雷高蒂（的想法）不同，对于我们来说，最能代表浦江新镇区域形态的建筑形态却是居住建筑。一般情况下，当我们看到筒子楼或者条形住宅我们就直接把它们和城市或者说一个特定区域的形态联系到了一起，这是因为中国城市的形态里，千篇一律的南北向建筑已经深入人心。而且浦江新镇中绝大部分的建筑都是住宅，整个设计方案也是在具体审计了居住社区的类型之后把它们一起放在基地上、放入到城市框架中去，这是另一个我们认为居住建筑更具代表性（更能代表整个区域形态）的原因。十分有趣的是，设计师对于所有的街区，依然采用了一致的体系结构，却发明了一种融合别墅、联排和公寓的更为灵活的建筑类型。（图11-12）根据功能设定的需要，建筑可以相应变化，比如设计有更多比重的别墅特质而可以作为别墅产品出售。这样的设计给人们一种居住为主的混合社区印象，同时也是一个现代的混合型中意新镇。这其实才是浦江新镇真正的区域形态。

8 奎多·莫泊格，维托里奥·格雷高蒂，《格雷高蒂国际建筑设计：城市景观的建筑》，纽约：里佐利出版社，2014年

9 薛求理，周鸣浩，《引入与适应：上海"一城九镇"的实践：格雷高蒂的浦江新城案例分析》，《国际城市设计》2007年第12期：21-40.页

Zhuo Cheng

On Tabula Rasa: A "Great Leap" in Urbanization

Fig.1 One City Nine Towns Plan
Source: Hartog, H. Den. *Shanghai New Towns : Searching for Community and Identity in a Sprawling Metropolis*. Rotterdam: 010 Publishers, 2010

图1 一城九镇计划规划图
来源：邓·哈尔托赫，《上海新城：追寻蔓延都市里的社区和身份》，鹿特丹：010出版社，2010

The Garden City and the One City, Nine Towns Initiative

A decade ago, as Shanghai's population approached 18 million and housing prices skyrocketed, the city decided to decentralize its development. Urban planners developed an initiative called One City, Nine Towns (Fig.1) – satellite suburbs would be built on farmland outside Shanghai to house one million people by 2020.[1] The One City, Nine Towns initiative as devised by the Shanghai municipal government was meant to save the metropolis from inevitable overpopulation. The plan was heavily influenced by Ebenezer Howard's Garden City (Fig.2), in which he suggested a multi-nodal approach to urban development with concentrated urban cores connected by public transportation lines, surrounded by unspoiled natural parks and agricultural zones; Shanghai's urban expansion plan can be seen as a development of this model, but with the additional stipulation of a specific architectural style for each of the new towns, creating beautiful, attractive alternatives to the downtown core. Each of the nine towns was themed with a style from a different geographical region. These developed towns were: Gaoqiao (Holland), Fengcheng (Spain), Pujiang (Italy), Anting (Germany), Songjiang (England), Luodian (Northern Europe), Fengjing (North America), Zhoujiajiao (traditional Chinese-style water town), and Zhoupu (Mixed Western).

Both the government and developers thought European themes would be attractive to Shanghai's new rich groups, but more than ten years after launching the project, the nine towns have been built to various levels of completion: some themed towns still remain empty, some have been canceled or put on hold, some are in progress, and yet others have stalled as victims of poor planning or political graft. All of the towns, says French architect Rémi Ferrand, who studied them as part of a book about the region's development, fit into Shanghai's landscape in different ways; the city, with its period of British and French occupation has always been regarded as a somewhat foreign place. Building these international "New Towns" is, in a way, *"like the continuation of a story."*[2]

OCT - Singular Developer

OCT Group is a large-scale, state-owned enterprise group engaged in cross-sector and cross-industry operation. It has fostered three major businesses in China, namely tourism and related cultural industry operations, real estate and hotel development, and operations and manufacture of electronic and packaged products. The tourism industry is its most influential main business and OCT has become the first brand in the travel industry in China. Depending on rich cultural tourism resources, the OCT culture

[1] http://www.smithsonianmag.com/people-places/shanghais-european-suburbs-428445/?no-ist.

[2] http://www.smithsonianmag.com/people-places/shanghais-european-suburbs-428445/?no-ist.

成卓　　　空白之上：城市化中的"大跃进"

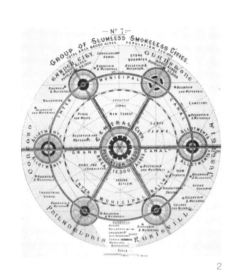

Fig.2 Ebenezer Howard Plan for Garden CitySource: Howard, Ebenezer, Thomas, Ray, and Potter, Stephen. *Garden Cities of Tomorrow*. New Rev. ed. Eastbourne, East Sussex: Attic Books, 1985

图2　埃比尼泽·霍华德田园城市规划图
来源：埃比尼泽·霍华德，雷·托马斯和史蒂芬波特·《明日的田园城市》，伊斯特本，东苏塞克斯：阁楼出版社，1895

"田园城市"和"一城九镇"计划

十年前，随着人口规模接近1800万和房价的快速增长，上海市决定疏散其中心区的发展。城市规划者制定了"一城九镇"的发展计划（图1），所谓的"一城九镇"计划，就是2020年以前，在郊区建立卫星城市以容纳一百万的人口。[1] 上海市政府设立"一城九镇"的目的是为了防止大都市中心区人口的过度膨胀。该计划很大程度上受到了埃比尼泽·霍华德提出的花园城市的影响（图2），他提出了城市多节点的发展方式，通过公共交通线路来连接高度集中的城市内核，以及未受破坏的自然公园和农业区环绕。这个城市的扩张计划可以看作是这种模式的发展，但是每一个新市镇有一个特定的建筑风格，创造出美丽而有吸引力的城市中心区域的替代品。每一个城镇会根据地理位置确定一种统一的建筑风格，高桥镇建成荷兰式现代化城镇，奉城镇建成西班牙风格小城，浦江以意大利式建筑为特色，安亭建成德国式小城，松江镇建成英国风格，罗店北欧风格，枫泾结合美国城镇风格，朱家角镇既凸现本土水乡古镇风貌，又有现代城镇的格调；周浦建成混合的西式风格。

无论是政府还是开发商都一致认为欧洲的主题将吸引上海的新富人群，但项目启动十余年后，九镇的建设各有不同的完成度：一些主题的城镇依然空空荡荡，有的已被取消或搁置，有的仍在建设当中，有些因为规划制定或者政治的原因停滞不前。如法国建筑师雷米克莱蒙费朗——他曾将这些城镇作为区域发展一书的研究对象——所说，这些城镇以不同方式契合上海整体的城市景观；这座曾经被英法占领的城市某种程度上一直被看作"外国城市"，建立这些国际化的新城，某种程度上是不是可以认为是"*曾经这个故事的延续*"？[2]

华侨城——单一的开发商

华侨城集团是一家从事跨区域和跨行业经营的大型国有企业集团。它在中国主要经营旅游及相关文化产业经营，房地产及酒店开发和经营以及电子产品制造三大主要业务。旅游业是最具影响力的主营业务，它已成为中国旅游行业的第一品

---

1 http://www.smithsonianmag.com/people-places/shanghais-european-suburbs-428445/?no-ist

2 http://www.smithsonianmag.com/people-places/shanghais-european-suburbs-428445/?no-ist

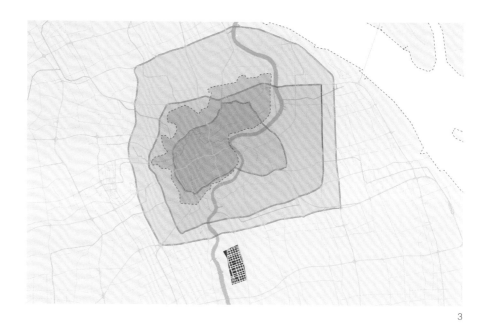

Fig.3 Locality of Pujiang New Town
图3 浦江新镇与上海市区的区位关系

industry, with its unique charm, is becoming the model of the industry. Some of its brands are well-known, such as Konka, Splendid China, Window of the World, Happy Valley Chain, and Portofino. It's the only Asian enterprise ranked in the quarter-finals of the Global Theme Park Group.[3]

In 2002, the Minhang District Government accepted funds from Shanghai Highpower-OCT Investment LTD for the development of Gregotti's Pujiang New Town project. Currently, a 2.6 square kilometer section in the north of the town has been built by the developer with the continuing involvement of Gregotti, as well as other Italian architects. The so-called "Phase 1" included the development of six blocks, and most of the available lands were sold to OCT. Even in China, one singular developer controlling the development of such a large area is rare, especially when the expertise of the developer is cultural and touristic. We can see a strong similarity between Pujiang New Town and OCT's development in Shenzhen, which is one of their most famous touristic projects. Basically, Pujiang is the downtown area; the development mode should be different for downtown residential areas and suburban cultural spots. What's more, a culture-related developer would certainly be subject to the resources within their network.

Locality and Connectivity

Unlike many of Shanghai's other suburban developments, Pujiang is only 16 kilometers from the city center, and can be easily reached in a reasonable amount of time by taking a ride down subway line 8 (Fig. 3-4). The convenient location provides great potential to sell the real estate properties due to its low transit cost to the city center and potential urban expansion. Initially, Shanghai's periphery was planned to become 24-hour cities by providing future job opportunities and services, as well as residential areas. The aim was to develop self-contained entities, which reduced pressure on Shanghai center, enabling the metropolitan area to accommodate future migration. The strategic plan aimed to avoid mistakes occurring in Western satellite cities, which often have become only residential 'sleeping towns' instead of a 'city' with various economic activities. The location of Pujiang, however, made it very likely to be developed as a bedroom community for people who work in the downtown but can't afford the housing prices there. The area therefore represents not a suburban district but rather an extension of the inner city. In addition, it brings speculation in real estate values; although most of the apartments are vacant, many of these apartments actually have been sold. Many properties are purchased as investments, as ways of storing money or as homes for future use.

3 http://www.chinaoct.com/category.aspx?NodeID=88.

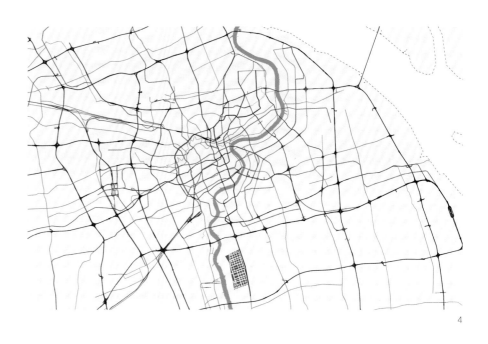

Fig.4 Connection of Pujiang New Town
图4 浦江新镇与上海市区的连接关系

牌。根据丰富的文化旅游资源优势，华侨城文化产业正在以其独特的魅力成为行业的典范，一些著名品牌是众所周知的，如康佳、锦绣中国、世界之窗、欢乐谷连锁、波托菲诺等。它是唯一一个全球主题公园八强的亚洲企业。[3]

2002年，闵行区政府接受了上海天祥华侨城投资有限公司对于格雷高蒂的设计方案开发的投标。目前，在该镇北部的2.6平方公里部分已被开发商天祥华侨城投资与格雷高蒂，以及其他意大利建筑师建成。第一期的开发包含的六个地块的发展，大部分可用土地都出售给了华侨城。即使在中国，一个单一的开发商来开发如此大面积的城市区域也是罕见的，尤其是这个开发商的专长是文化和旅游业。我们可以看到，他们在深圳最有名的旅游项目之一华侨城和浦江新镇之间具有高度相似性。但是浦江本质上是一个城镇的中心区域，其开发模式与位于郊区的文化场所之间肯定是有所不同的，另外文化产业相关的开发商在开发城镇中心住宅和商业区域时肯定会受到其资源的限制。

地域与连接

不同于"一城九镇"的其他城镇，浦江距离上海市中心仅16公里，并且可以通过采取搭地铁8号线在合理时间内轻松抵达（图3-4）。其方便快捷的位置由于其未来可能的城市扩张和相对较低的出行成本对房地产市场的发展提供了巨大的潜力。最初的计划是通过提供就业机会和服务以及住宅区，使上海的周边地区成为24小时的城市。这样做的目的是发展成一个自成一体的城市，它能减少上海中心的压力，使大都市区能容纳未来的人口迁移。该战略计划的目的旨在避免西方城市发展中出现的错误，就是往往成为一个只具有"睡觉"功能的区域，而不是有各种丰富经济活动的"城市"。然而由于浦江的位置，它很容易发展成一个仅仅为了在中心城区工作却无法负担高房价的人群居住的城镇。因此，其代表的并不是一个郊区自成一体的卫星城市，而是城市内核的延伸。此外，它带来了房地产价值的炒作，虽然大部分公寓都空置，但实际上都已经售出。很多房地产是出于投资购买的，作为一种存钱的方式或者作为以后的住宅。

[3] http://www.chinaoct.com/category.aspx?NodeID=88

In Ebenezer Howard's Garden City, the network nodes of a multi-center ecological metropolis were tied together into a whole by public transportation links, and it was not expected that any one node could survive separately. But even with the subway line 8 connecting it to Shanghai center, Pujiang and other towns still lack any connection with each other, requiring a car, a circuitous bus journey, or a long trip on the metro and a taxi ride to reach. With better transportation connections within the network of suburban towns, not just Pujiang but all the towns, are more likely to grow into thriving subcenters of Shanghai's rapidly developing suburbs.

Ghost Town Phenomenon

Over the past few years, China's urban population has grown rapidly. Today the urban population makes up about 54% of the total population. The urban population needs housing, but a lot of the construction in China comes in anticipation of migration. Many developments are ghost towns, which are new cities that have yet to come to life due to inappropriate planning and development. Pujiang is actually a typical case resulting from similar conditions.

In China, it's common policy that local governments use publically-owned land for speculation. China's fiscal policy requires local municipalities to comply with this broader urbanization plan. Local municipalities must fend for 80 percent of their expenses while only receiving 40 percent of the country's tax revenue. Land sales make up much of the difference, resulting in a buy low, sell high scheme, as municipalities buy up cheap rural land, re-designate it as urban, and then re-sell it at the high urban construction land rate – pocketing the difference.[4] The first phase of any project usually serves as an attractor for further development steps, which is the intention of any local government in China, and in fact a major source of revenue is to gain maximum profit from land speculation. In most cases, from the perspective of government, the success of a project is evaluated mainly by the profit it produces; they deliver the land for housing to gain revenue. As a result, housing in many new cities in China is built prior to infrastructure like hospitals, schools, places to work, and, often, good public transportation networks to generate revenue.

Pujiang is a typical case of developing housing prior to infrastructure and public facilities. Residential program occupies a much larger area than other programs, which is obviously insufficient to support the development of the town; therefore the town is inevitably developed as a bedroom community with imbalanced program uses. There are

[4] http://blogs.reuters.com/great-debate/2015/04/21/the-myth-of-chinas-ghost-cities/.

按照霍华德花园城市的构想，多中心大都市生态的网络节点是由公共交通线路连接成一个整体，任何一个节点无法单独生存。不过，即使有8号线连接上海中心，浦江与其他城镇仍然缺乏彼此间相互的连接，需要迂回的公交线路，或长时间的地铁转换，乘坐出租车才可到达。如果郊区城镇网络之间能有更发达的交通网络，不仅仅是浦江，而是所有网络内的郊区城镇都有可能成长为上海快速发展的郊区的一个蓬勃发展的城市副中心。

鬼城现象

在过去的几年中，中国的城镇人口迅速增长。如今，城市人口构成了总人口的约54％。那些城市的人口需要住房，但中国很多城市的建设是基于对农村人口转移的推测。我们看到很多鬼城由于不恰当的规划和发展而缺乏相应的活力，而浦江实际上是相似情况下的一个典型案例。

在中国，地方政府使用出租国有土地获利是普遍政策。中国的财政政策要求地方政府协调遵守这一城市化计划。当地市政府必须承担80％的花费，而只能收到税收收入的40％。卖地收入弥补了收支的巨大差异，造成了土地低买高卖。地方政府便宜收购农村土地，重新规划它为城镇，然后以城镇建设用地转售——赚取差价。[4] 项目的第一阶段通常被用来吸引进一步的投资，土地收入作为财政收入的主要来源，政府的意图实际上是为了获得土地投机的利润最大化。在大多数情况下，从政府的角度来说，一个项目的成功主要是通过它产生的利润来评估，所以他们经常首先交付住房用地来获得收入。其结果就是，为了产生财政收入，中国许多新兴城市的住宅建设往往早于基础设施，如医院、学校、公共交通网络等。

浦江就是一个房屋开发先于基础设施和公共设施的典型案例。住宅占有比城市其他功能更大的区域显然不足以支撑城镇的发展。因此，在城镇内部居民没有足够的就业机会，该镇将不可避免地发展为各城市功能不平衡的"睡城"。像许多华侨城的社区一样，家乐福是目前在浦江镇唯一的商业设施，但超市显然难以发展

[4] http://blogs.reuters.com/great-debate/2015/04/21/the-myth-of-chinas-ghost-cities/

Fig.5 Exterior View along the River on Fitness Center and Townhouse
Fig.6 Night View of Multi-Storey Residence
Fig.7 Courtyard View of Clustering Residence

图5 健康中心及联排住宅沿河外景
图6 多层住宅夜景
图7 独院住宅组团内景

not enough jobs offered in the town to convince residents to settle down. The Carrefour, which exists in many OCT communities, is currently the only commercial facility in Pujiang, but a supermarket is not likely to develop into a commercial hub that attracts a population influx. After more than ten years of development, there is a distinct lack of anything in this massive community (Fig. 5-7). Not only does Pujiang's Italian town seem uninhabited, it is pretty much uninhabitable.[5] The lack of variety in terms of program results in the lack of vitality in the community. When large numbers of people move into a new area, they need to be provided with public services like healthcare and education. Therefore, a population carries a price tag and there is often an extended period of time between when cities appear completed and when they are actually prepared to sustain a full-scale population.[6] But this creates a contradiction to the ghost city phenomenon: few people are going to move into a place that lacks adequate public facilities for urban life, but from the developer's perspective, what would be the point of providing these public facilities until there are people?

In addition, as discussed above, many of the apartments in Pujiang actually have private owners. As is the case with many of China's other ghost cities, many of the properties are purchased as investments, as ways of making money, and as homes for future use. So many, if not all, of the properties sell, but people don't intend to actually live in them – at least not any time soon. That's also the reason for such a high vacancy rate in the community. This has had a severe effect on placemaking and community building in many of the new towns. As almost no residents have yet moved into the new towns; retail and public services are not feasible and therefore also left empty. Speculation has become a vicious spiral keeping the ghost towns uninhabited.

In China's rapid urbanization there is hardly a single new urban development in the country that has yet gone over its estimated timeline for completion and vitalization, and as the cities are expanding, the ghost town is mostly a phase of development. China's new cities and districts are long-term endeavors. Most large, new urban developments in China eventually move through the ghost town phase and become vitalized with businesses and a population. So the ghost city label at this point is premature: most are still works in progress. When infrastructure is built, shopping malls open, small-scale public facilities are provided, and places where residents can work are created, the cities will come to life.

5 A Journey To China's Italian Ghost Town, Wade Shepard. Source: http://www.vagabondjourney.com/china-ghost-town-italy-style-pujiang-shanghai/.

6 http://blogs.reuters.com/great-debate/2015/04/21/the-myth-of-chinas-ghost-cities/.

成为区域的商业中心来吸引人口流入。经过十余年的发展，这个庞大的社区明显缺乏与城市相关的东西（图5-7）。浦江的意大利小镇似乎无人居住，实际上它几乎也是无法居住的。[5] 城市功能多样性上的缺乏导致了社区活力的缺乏。当大量的人口进入一个新的城市区域，他们需要医疗和教育等公共服务。因此，人口是有价格标签而且当城市建设看似完成时和它们实际上能够承受大量人口居住之间有一个较长的时间间隔。[6] 但是，这就造成鬼城现象的矛盾性，很少人会选择住进一个缺乏足够公共设施和城市生活的地方，但是从开发者的角度来看，在有大量人口进入以前提供公共服务和基础设施的意义又何在呢？

此外，正如我们上面所讨论的，许多浦江的公寓实际上都由私人持有。如同中国其他地方的鬼城现象，许多房地产，如果不是全部的话，属于投资性购买或者作为家庭未来使用的储备。但是人们不打算生活在其中，至少近期内不会，这也是造成许多社区空置率如此之高的原因。这已经对许多新市镇的城镇建设和社区营造造成了严重影响。由于几乎没有居民迁入新市镇，零售和公共服务的建设在经济上几乎不可行，因此这些社区也是空的。房地产投机造成了鬼城现象并且已经成为一个恶性循环。

在中国的快速城市化进程当中，几乎没有一个城市的发展经历了完整的时间线。随着城市发展的不断扩大，鬼城成了发展的一个阶段。中国的新城建设和区域发展是长期的。在中国大多数大型新城市发展最终通过了鬼城这个阶段并成为集商业和人口活力的都市。所以，鬼市在这一点上是城市发展不成熟的标签：大多数仍是在进行中的工程。当基础设施建设完成，商业设施入住，小规模的公共设施和工作机会得到提供，这个城市将会"复活"。

5 韦德·谢帕德，《中国意大利鬼镇之旅》
资料来源：http://www.vagabondjourney.com/china-ghost-town-italy-style-pujiang-shanghai/

6 http://blogs.reuters.com/great-debate/2015/04/21/the-myth-of-chinas-ghost-cities/

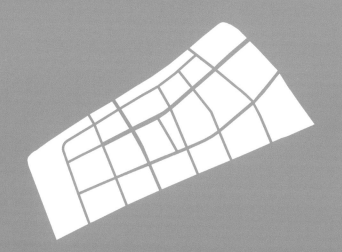

Cultural Multiplicity
文化多元

Biyun International Community
碧云国际社区

Developing semi-urban fields to international communities
发展半城市化区域为国际社区

From International Community to Global City

A Brief Study of Social, Cultural, and Spatial Urban Mixing in China based on the Biyun International Community

Ruoyun Xu

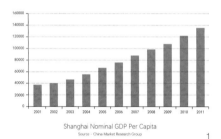

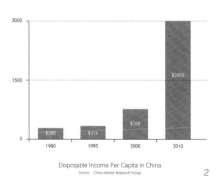

Fig.1 2001-2011 Chinese National Economic Boost
Fig.2 1980-2010 Disposable Income Per Capita in China

图1 2001-2011 上海人均GDP
图2 1980-2010 中国人均可支配收入

The Biyun International Community has been one of the most successful residential clusters originally planned and designed for international residents in Shanghai over the course of the last twenty years. While it's evident that we should acknowledge its success, we should also examine the case of Biyun considering that both the community itself and the city of Shanghai have experienced drastic changes during the two decades. It would be fantastically interesting to rethink Biyun and the whole variety of political, economic, cultural, and social issues it entails.

CHANGES

The substantial changes in Shanghai during the last two decades include primarily three aspects: the boost of national and local economy, the transformation of residential demographics and the evolution of urban structure.

Economic Boost - During the last two decades, especially at the beginning of the 21st century, the rapid growth of the Chinese economy enabled a boost of GDP per capita (Fig.1) and disposable income (Fig.2), which empowered Chinese people to be more capable of affording a better domestic life. On the other hand, although the income increased drastically on average, the gap between rich and poor became even more magnified (Fig.3), with the richest community having their living conditions improved significantly while the poorest stayed where they were decades ago.

Demographic Composition - Economic development catalyzed the transformation of demographics. Biyun was originally designed as a community for foreign recruits working and living in Shanghai, most of whom are temporary residents. The community has since been occupied by a growing percentage of Chinese residents working for the foreign corporations, or others in the area, which transforms the lifestyle of the community to be more diverse and oriented toward integration. For the first time in recent decades, the number of foreign residents in Shanghai was reported to be in decline in 2014, the reason arguably being that, the slight slow-down of the Chinese economy triggered an adjustment of investment strategy of foreign corporations. In the foreseeable future, therefore, there will be less foreigners coming to Shanghai and staying for business purposes while more will come for non-business and permanent purposes, which indicates a new pattern of cultural exchange between Chinese and foreign populations.

Urban Structure - The boost of the economy and the transformation of demographics influences the evolution of urban structures, especially transit systems, largely increasing the level of mobility and altering the scale of time and space. Biyun, a good illustration of urban expansion, has been redefined by the ring road system. The construction of both the inner and outer ring roads of Shanghai was started in 1993, and was completed by 2000 and 2003 respectively, reframing the urban structure and transit modes and redefining the spatial interrelation between the center and the periphery. The Biyun

从国际社区到全球都市

基于碧云国际社区浅谈中国城市的社会、文化与空间融合

徐若云

在过去20年中,上海碧云国际社区已经成为上海为外国居民规划设计的居住社区中最成功的案例之一。尽管碧云的巨大成功是显而易见的,但我们仍应看到,不管是上海还是碧云社区自身,都在这20年中经历了翻天覆地的变化,而这种变化也将极大地影响碧云社区。在这种语境下,探讨碧云社区与牵涉其中的政治、经济、文化、社会议题将会十分有趣。

变革

对于上海20年间的剧烈变化,笔者将重点关注三个方面:中国与上海的高速经济发展,上海城市居民的人口组成变化,以及上海城市结构的演变。

经济水平——过去20年中,尤其是新世纪头十年,中国经济保持高速发展,国民生产总值(图1)与人均可支配收入(图2)大幅度增加,越来越多的中国人有能力购买面积更大的住房,获得更好的居住条件。然而,虽然人均可支配收入增长迅速,我们也需要看到,不同群体的人均收入差距也在同时增大(图3)。随着富裕群体的居住环境日新月异,而许多贫困群体的居住条件并未获得较大改善,不同收入群体所代表的社会阶层矛盾日益激化。

人口组成——近年国民经济水平的提高,加快了上海城市居住人口组成的变化。碧云国际社区最初是为金桥开发区外国企业中的外籍员工所设计的,早期社区居民多数是出于公务需要在上海短期居留。近年来,一方面大量中国企业实力增长迅速,开始进驻开发区,另一方面外国企业中中国员工数量增加,碧云国际社区的中国居民比重也逐渐增加,极大地增加了居住环境与生活方式的多样性,也使得社区融合成为一种不可阻挡的发展趋势。2014年,上海的外国居民人数近年来首次下降,其原因或许是中国经济增速放缓所导致的外国企业市场策略变化。因此,我们可以推测,在可以预见的未来,外籍人士出于公务需要在上海短期居留的情况会逐渐减少,而出于个人生活的意愿在上海长期居留的情况会增加。这预示着中外居民之间将出现新的文化交流模式,同时对国际社区的空间也提出了新的需求。

城市结构——经济水平的提高和人口组成的变化同时影响着城市结构的演变,其中最主要的演变之一是交通系统带来的城市扩张,即其对城市尺度与城市机动性能的影响,而碧云国际社区恰好就是这种城市演变的充分证明。上海内环路与外

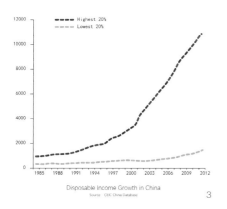

Fig.3 1986-2012 Disposable Income Per Capita in China - Comparison between the Highest 20% and the Lowest 20%

图3 1986-2012 中国人均可支配收入贫富差异

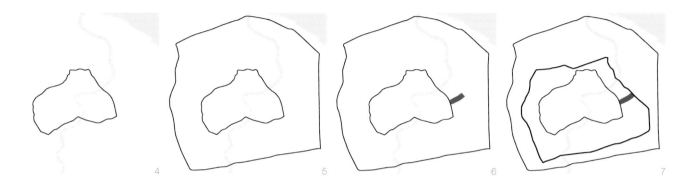

4　5　6　7

The Spatial Relationship between Biyun Community and Shanghai Ring-Road System

Fig.4 Shanghai Inner-Ring construction started in 1993 and finished in 2000, as the first ring road of the city, has altered the urban scale and transit pattern

Fig.5 Shanghai Outer-Ring construction started in 1993 and finished in 2003, completed the double ring road system and redefined the relation of urban center and periphery

Fig.6 Biyun International Community construction started in 1997, first as an attached part of industry in the suburb between the inner and outer ring road, was far away from the dense urban area

Fig.7 Shanghai Middle-Ring construction started in 2000 and finished in 2015, and transformed Biyun from an attached part of industry to an integral part of city in between the inner and middle ring roads in the dense urban area

上海城市环路结构与碧云社区的空间关系

图4 上海内环路1993年开工，2000年完工，作为首个城市环路，改变了城市空间尺度与交通模式。

图5 上海外环路1993年开工，2003年完工，与内环路形成了双环结构，重新定义了城市中心与城市近郊的空间关系。

图6 碧云国际社区1997年开工，当时社区作为工业区配套设施，处于内环与外环之间的近郊地区，远离密集城市环境。

图7 上海中环路2000年开工，2015年完工。碧云社区处于内环与中环之间，周边城市密度迅速提升，社区定位从工业区配套转变为城市有机组成部分。

International Community, which was originally intended for the residential part of the Jinqiao Export Processing Zone with construction starting in 1997, located itself in the low-density suburban area between the inner and outer ring roads at that moment. Since the construction of the middle ring road was started in 2000 and completed in 2015, the density of that area has been increasing constantly, transforming Biyun from a residential component of a suburban industrial zone to an integral part of a dense urban center (Fig. 4-7).

In this sense, the Biyun International Community, after claiming its success for the past twenty years, has started to be challenged by more divergent income groups, more diverse cultural backgrounds and more integrated urban structures, which might suggest certain changes in the community to be offered. I intend to reexamine the case of Biyun through context observation, case investigation, and scenario speculation in aspects of policy, economy, and culture, as well as urban planning and design, so as to rethink the opportunities of Chinese urban mixing in the future.

OBSERVATIONS

From a conventional perspective, a series of features of the Biyun International Community can be interpreted as both advantages and disadvantages. The location of the community, which is in the midst of Lujiazui CBD, Jinqiao Export Processing Zone, and Zhangjiang Innopark, not only identifies the residential demographics as primarily white-collar with exclusively high incomes, but also encourages private vehicle ownership over public transit coverage and the development of super blocks at an unfriendly scale for pedestrians (Fig. 8). The relatively organic mixing of residential and commercial programs promotes the hybridized urban life within the community, while halting the connection between inside and outside. The international schools offer an excellent amenity for the foreign residents, but, at the same time, intensify the segregation of Chinese and foreign populations. The gated and fenced communities, while in a sense increasing the level of security, isolate the public space and private space with no transitions, especially for pedestrians. All the features above, while providing certain advantages, weave together and cocoon Biyun into a self-sufficient island floating in the city.

Within the discourse of urban design and architecture, it's a common belief that the city should be an organic whole with a good mixing of programs, demographics, and scales. This mindset, instigated by the New Urbanism conventions prevailing in 1980s consumerist American cities, has prevented contemporary Chinese public opinion from realizing that China is still in the middle of its long economic journey and that the economy remains underdeveloped and the gap between rich and poor keeps growing. The fierce conflicts between different social classes will still be irreconcilable if we simply "mix" within the scope of urban and architectural design, without considering social justice.

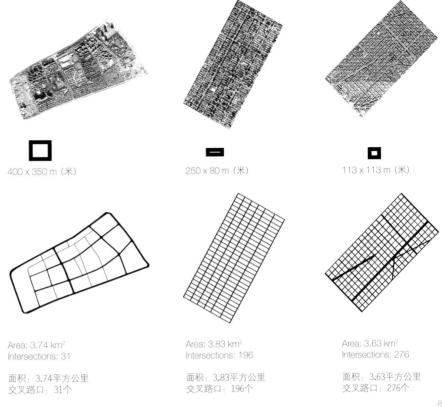

Fig.8 Comparison on Urban Density of Residential Areas - Biyun International Community, Shanghai - Upper East, Manhattan, New York - Eixample District, Barcelona

图8 居住密度对比：上海碧云社区/纽约上东区/巴塞罗那扩展区

环路于1993年同时开工，分别于2000年与2003年竣工，其形成的环路交通系统从根本上改变了城市结构与交通模式，重新定义了城市中心与郊区的空间关系。于1997年开工的碧云国际社区，最早被定位为金桥国际开发区的配套居住设施，坐落于内环与外环之间，在当时属于低密度城郊地区。随着中环路于2000年开工，2015年竣工，碧云国际社区周边城市密度持续提高，将碧云社区从一个城郊工业园区的配套居住设施转化为一个密集城市中心的有机组成部分（图4-7）。

从这个意义上，碧云国际社区在经历了20年的成功之后，开始受到各种新的挑战：更悬殊的收入群体，更多样的文化背景，以及更复杂的城市结构。面对这些新的挑战，碧云国际社区需要从功能定位到空间设计都作出相应的改变。本文将通过现状调研、案例分析与未来展望三个部分，从政策、经济、文化以及城市规划设计等角度对碧云国际社区进行重新思考，并以此为基础审视中国未来城市融合的可能性。

现场观察

从城市与建筑的传统视角看，碧云国际社区的一系列特征既是经济发展的机会，又是文化融合的挑战。第一，碧云社区位于陆家嘴CBD中心区、金桥开发区和张江技术园之间，不仅将住户定位于高收入白领人群，同时也提高了私人汽车的利用率，阻碍了该区域公共交通的发展，因而导致街区尺度过大，对步行者而言不够舒适（图8）。第二，相对而言，社区中的居住与商业功能有一定的混合，激发了社区内城市公共生活的活力，但同时也减少了社区内外的交流。第三，社区内的国际学校为外国居民提供了较为便捷的生活配套设施，但同时也加剧了中国居民与外国居民的隔离。第四，社区完全被大门与围墙包围，虽然在一定程度上增加了安全感，但也阻隔了公共空间与私密空间的流通性。上述特征共同影响着碧云社区，并将其逐渐转化为一座自给自足的城市孤岛。

Everyone loves beautiful gardens, but not everyone who paid for the garden is willing to share it with someone who didn't. If we simply demolish the gates and fences and open up the community, or mix the housing units with low-income and high-income residents, or force Chinese and foreign residents to follow the same lifestyle, we are probably pleasing the outsider, the low-income groups or Chinese residents and upsetting the insider, the high-income groups or foreign residents. How can we fulfill the aspirations of some groups by sacrificing those of other groups? When we claim that we design for the public, who is the public? Only the poor? Are the rich not part of the public? Is it reasonable that we neglect the rich because they already live a better life? Is "mixing" feasible in this context? If so, how should we achieve it? And to what extent? If not, how should we approach the problem of Biyun as an "urban island"?

IMPLICATIONS

While we are rethinking the case of Biyun, it is also necessary that we expand our horizons to scrutinize existing cases which tackle the issue of urban mixing. As tremendously as the cases might vary in terms of their contexts, they are still eligible to offer referential thoughts for Biyun.

Between the Chinese and the foreigners - Wangjing (Fig. 9), one of the largest Korean expatriate communities in China, is located in the northwest part of Beijing. It has been attractive to Korean populations in Beijing since it was completed due to its proximity to the Korean embassy, the airport, and the business district where Korean companies aggregate, and the relatively low rents compared to other nearby communities. Although originally intended as the first high-end residential community, exclusively targeted at the middle class, it now has become a social mixing hub in terms of its diverse classes, ethnicities, and cultures.

Korean residents in Wangjing have been purchasing units, rather than renting, because they would like to stay in Beijing with these outstandingly convenient and affordable amenities. Many even purchased multiple units and profit from renting them out, which sometimes covers their own mortgage, since the companies and universities nearby ensure relatively stable sources of tenants. With this tendency of purchasing units, most Korean residents migrate to Beijing with their whole family, bringing a Korean lifestyle with authentic Korean shops and restaurants and keeping the Korean cultural atmosphere. Because of the influence from the Korean Waves, Chinese people are willing to come to Wangjing for residency or consumption, which largely promotes the Wangjing Community in terms of economy, culture and urban transportation.

As the contact zone of Chinese and Korean cultures, Wangjing is confronted with divergent social classes, income statuses, religious beliefs and cultural traditions. Although interested, even amazed, by the cultures of one another, Chinese and Korean

Fig.9 Wangjing Community, Beijing, China
Source:https://antony2012.tuchong.com/12601300/

图9 北京望京社区鸟瞰
来源: https://antony2012.tuchong.com/12601300/

从城市设计的角度看，城市应该是一个混合多种功能、人群与尺度的有机整体。这种来源于80年代广泛流行于美国的新城市主义思想，在很长一段时间里都对中国城市的发展方向产生误导，使中国的城市管理者与规划者们忽略了一个重要的事实：中国的人均经济、文化与教育水平仍处于初级阶段，社会各阶层之间的差距还在不断扩大。因此，如果我们不考虑社会公正等因素，而是简单地在城市设计中推行"混合"的设计思想，不同社会阶层之间的矛盾仍然无法调和。

每个人都希望拥有美丽的花园，但并非每个人都愿意将自己花钱买的花园拿出来与他人分享。如果我们简单地拆除封闭社区的大门和围墙从而将社区开放，或者将低收入居民与高收入居民简单地在一栋住宅中混合，或者为中国居民与外国居民安排同样的生活环境与生活方式，一部分人也许会受益，但另一部分人则会因此不满。我们能否为满足一部分人的愿望而牺牲另一部分人的利益？当我们自称为公众而设计时，"公众"指的是谁？只有低收入阶层？高收入阶层究竟是否是"公众"的一部分？因为高收入者生活水平较高，我们在设计时就可以不考虑他们的需求？在这个语境下，"混合"是否可行？如果可行，我们如何操作？在何种语境下操作？如果不可行，我们如何解决碧云社区的城市孤岛问题？

启示

当我们思考碧云社区的时候，我们同样也应该扩展自己的视野，学习现有的处理城市混合问题的案例。虽然这些案例的背景可能大不相同，但我们仍能从其中找到一些具有参考价值的启示。

中国居民与外国居民——位于北京东北部的望京社区（图9）是中国最大的韩国移民聚居区之一。望京形成与繁荣的原因可以归结为三点：第一，靠近韩国大使馆与北京国际机场；第二，靠近韩国公司驻京办事处聚集地；第三，相对其他社区租金较低。虽然最初市政府对望京的期望是北京首个面向中产阶级的高端居住区，但目前望京已经成为不同阶层、国籍、民族与文化高度融合的混合型社区。

与碧云不同的是，望京的外国居民大部分购置房产，而非短期租赁，因为他们喜欢北京便捷的生活和低廉的物价。一些居民甚至购买了多套住宅，凭借周边密集的公司或高校带来的稳定客源来以房养房。这种购置房产的偏好使得韩国移民大多将一家人都接到北京居住，同时也在住处附近经营韩国商店或餐厅，将原汁原

communities still have suffered from the discrepancy of lifestyles and customs. For example, Chinese residents use to organize noisy square dances and cook Chinese food with oil fumes, both of which are annoying for Koreans. Korean residents, on the other hand, used to hang out till midnight, usually getting drunk and riding powerful motorcycles, which goes against Chinese lifestyles.

To solve this problem, the Wangjing Community started to employ Korean-Chinese people, who are Chinese citizens fluent in both languages, as community coordinators to enhance the communication and interaction between Chinese and Korean residents. The duties of these coordinators include translating Korean community publications and collecting information about Korean community activities for Chinese police, so that different lifestyles and aspirations can be mediated by police and residents, eventually leading to self-disciplined community management.

However, conflicts keep emerging all the time in the Wangjing Community. Byoungyoun Joo argues in his dissertation that there are three approaches to improve this situation. First, complete national legislation on issues of foreign communities in China, taking full consideration on different customs and lifestyles, so community management can follow a consolidated system of rules rather than loosely organized moral agreements. Second, establish local policies on community management to regulate the behaviors of residents, with a well-organized public participation mechanism to empower the residents toward a more self-disciplined inhabitation. Third, encourage bilateral cultural recognition based on the various similarities across China and Korea, so people will understand each other in a more insightful way.[1]

Between the rich and the poor - AVA High Line, a luxury rental at 525 West 28th Street, Manhattan, New York, offers 142 affordable units apart from its 710 market-rate apartments (Fig. 10). New York has been advocating for this type of mixed-income housing project for the last 20 years, which has lead to more than 100 mixed-income residential communities. In these communities, most apartments remain market-rate units, some even luxurious, while others, always around 20% of the total number of all units, are designated for low-income population (Fig.11). The government sets considerable incentives by reducing taxes or approving more construction or enabling more mortgages, so that private developers are more willing to designate a portion of their units to accommodate low-income residents.

[1] Byoungyoun, Joo, The status of Korean Social Intergration in Beijing - A case study of Wangjing "Korean Town", Shandong University, 2013.

Fig.10 AVA High Line Mixed-Income Housing, New York, USA
Source: http://observer.com/2014/06/ava-high-line-doesnt-want-your-security-deposit-she-just-wants-to-be-your-friend/

图10 美国纽约曼哈顿高线AVA混合收入住宅
来源：http://observer.com/2014/06/ava-high-line-doesnt-want-your-security-deposit-she-just-wants-to-be-your-friend/

味的韩国生活方式和文化氛围带到北京。另一方面，由于"韩流"在当代中国流行文化与传媒产业中的盛行，北京居民越来越愿意来到望京消费或居住，更多地接触韩国文化，这也在一定程度上促进了望京社区经济、文化与交通的发展。

望京社区中，存在着完全不同的社会阶层、收入状况和文化传统，其中冲突最激烈的当属中韩之间不同的生活方式。虽然在很多情况下中韩居民都对彼此的文化有着浓厚的兴趣，但日常生活的矛盾仍然处处可见。例如，中国居民经常组织广场舞并制造噪音，或在烹调时产生油烟，韩国居民对此常有不满；而另一方面，韩国居民深夜大声喧哗，或醉酒驾驶摩托车，也让中国居民多有抱怨。

为了解决这些问题，望京社区开始雇佣朝鲜族管理员，希望利用其语言优势来居中协调，提升中韩居民之间的沟通交流。这些朝鲜族管理员主要负责为中国民警翻译韩国居民的社区出版物和收集韩国居民社区公共活动的信息，从而方便民警和居民协调不同的生活方式与诉求，引导社区自治的管理模式。

尽管如此，矛盾与冲突在望京社区中仍然存在。韩国学者朱秉渊提到解决望京社区管理问题的三种方法：第一，完善针对外国社区的法制体系，全面考虑不同的传统习俗与生活方式，从而使社区管理可以有法可依；第二，建立社区管理制度，用更多的公众参与来规范居民行为，增强居民对社区的责任感和归属感，提高社区的自律性；第三，组织文化交流活动，增加中韩居民对彼此传统风俗习惯的了解与认识，促进不同国家之间的文化认同。[1]

高收入群体与低收入群体——高线AVA（AVA High Line），是一座位于纽约市曼哈顿区的高档公寓楼，其在提供710套市价单元的同时还提供了142套"可可支付单元"，即经济适用房（图10）。近20年来，纽约一直推行此类混合收入住宅项目，至今已有一百多个项目成功运行。大部分公寓单元仍然以市场价甚至更高价格出售或出租，而其他部分——通常约占总单元数的20%左右——只能出租给低收入人群（图11）。纽约市政府建立了一套奖励机制，通过减免税务或允许更大规模的开发等措施为开发商提供可观的补偿，因此开发商就更加愿意按照政策为城市提供更多的低收入住宅。

[1] 朱秉渊，《在京韩国人及其社会融合状况——以望京社区为例》，山东大学，2013

A Dialogue between the Local and the Global

A Study of Architectural Language as the Signifier of the Conflict between Urban Expansion and the Preservation of Local Identity based on the Case of Biyun

Huopu Zhang

Cities, as extensive entities that comprise a large amount of parts of varied scales and diverse natures, when viewed through scopes of different dimensions, provide us with different focal points into the matters of urban momentum and the socioeconomic factors that lie beneath them. All these various scopes can actually constitute a continuous spectrum of views that are related to all layers of the elements and issues of interest within a city, and, in return, each of them can be taken from such a spectrum to address a certain subject that is best represented in such a scale.

In particular, when looking into the case of the Biyun neighborhoods in relation to their context, there are two scopes that stand far from each other on the spectrum that are highly representative and useful – the city as a whole, which is equivalent here to the city of Shanghai, and the scale of a building block, which can be any one of the blocks lying within the boundary of Biyun. Any specific architectural features, as far as the matter that they signify is concerned, can reveal to a certain degree the motives and the influence from socio-economic factors in different scales of the urban context. On the other hand, when these factors that are driven by different layers of information within the city, that are in conflict, the reconfiguration of the architectural elements that stand as their signifiers can help re-interpret the relationship between the parties in conflict and would contribute to mediating their needs. In my opinion, the pressure faced by the Biyun neighborhoods nowadays, both in relation to the change in climate in economics and the Chinese society and to the ever-accelerating pace of urban expansion of the city of Shanghai, is worth looking into through these different lenses. By analyzing the conflicts of interest in Shanghai as a whole and the desire of Biyun as a locality that is inevitably involved in the former as the climate changes, as well as the architectural factors that are related to these varied interests, we might be able to draw out some conclusions and possible solutions that can help ease the tension that is currently present.

As the study of the gradual development of the Biyun neighborhood and the recent change in socio-economic situations in and around Biyun has noted, there is an ongoing back and forth conflict between the interest in addressing the neighborhood's unique property - its identity as a foreign enclave - and the city's desire to absorb the neighborhood, exacerbated by the pressure of expansion it has always been facing. Both sides of the conflict show a strong relationship with the number of people, or, in socio-economic terminology, the demographics. Thus, the fluctuation of the influence of the two sides in the conflict can actually be interpreted as driven by the shifting of Shanghai's demographics and the related change of climate in the city's economy.

As we look into Biyun's earlier development and expansion, we can see that it benefited both from Shanghai's policy of encouraging foreign investments in the Pudong New District, and from the fact that Biyun's location was for a long time quite distant from Shanghai's city core. Development in Biyun was pushed forward by the city's interest in building up foreign communities to promote its international influence, as well as the strong global economy at the time, which helped the growth of foreign firms in the

局部与全局的对话

基于碧云社区案例分析浅谈城市扩张与社区个性之间的冲突

张霍普

城市作为一个由大量不同尺度及性质的部分所组成的整体，能够在不同的观察焦点下，反映出蕴含在其不同尺度的分辨率下不同层次的社会经济因子和城市要素。所有的这些不同的观察层次，实际上给我们提供了一系列连续缩放镜头，而城市里许多无法在同一层面进行观察与分析的要素，则分别在这些不同分辨率的镜头里展开。另一方面，某些要素在特定观察层面下则能得到最好的呈现。

在这样的情况下，当我们着手调研上海碧云社区的案例，尤其是在城市的国际化与本土化的相互关系这个涵盖许多不同层次的议题上，通过两个彼此差别较大的、具有代表性的观察范围进行研究则相当有意义。这两个不同的观察范围分别从上海城市整体的角度以及碧云社区内部任意一个典型街区的范围出发，前者着重城市发展的有机性与整体性，后者则充分放大社区建筑空间物理层面的特性。不同观察层次下的所有建筑特性都不同程度地能够反映出该观察层次下比较显要的社会经济因子。而另一方面，当这些在不同层次下具有不同程度的代表性因子之间产生矛盾时，与这些影响因子与要素相关联的具体建筑形式的重整，也能在反映出冲突各方的内部关系的同时增进各方利益与兴趣的协调。在当下，碧云社区所面对的源于中国经济与社会变革以及上海不断提速的城市扩张进程的压力，非常值得透过不同的视角范围进行研究。通过分析上海这个城市整体的发展趋势以及碧云作为一个社区与其自身的利益追求之间的矛盾，以及这二者相对应的建筑层面的特性，我们应该能够得出一些缓解当下的矛盾局面的结论。

根据碧云社区从诞生到现在的发展历程，以及与之相伴的上海社会经济发展的历程来看，碧云其实不断地在两股作用力的驱使下进退，一方是碧云自身作为上海一个外国人社区、高端社区，而与其周边环境区分开来的独特性，另一方则是上海城市整体的发展定位于不同发展阶段下的发展目标。在过去，当二者相协调，或者说是上海本身鼓励在浦东远离上海市区的地方培育一个能够吸引跨国公司管理人员居住的精品社区的时候，碧云的发展是迅速的。而在目前的经济社会条件下，由于上海整体国际化水平的提升，个体社区的国际化定位遭到弱化，同时上海城市发展的空间向郊区推进，两方力量便陷入了冲突。往更深入的角度看，这两股作用力背后都与人的因素紧密相关。不同时期不同的人口构成以及变化趋势，在很大程度上影响了碧云社区作为一个个体与上海作为一个整体的发展气候。

具体来说，在碧云早期孕育以及扩张的阶段，我们可以清楚地看到其飞速的发展成熟得益于上海鼓励外资落地浦东新区的政策。另外，碧云本身所处的地理位置也与上海市中心有一定空间距离，不会过早与城市扩张的步伐发生正面冲突。在

Fig.1 Aerial Image of Biyun International Community
图1 碧云国际社区鸟瞰

nearby Zhangjiang Economic Zone. Given these contingencies, it is not hard to see how the situation would change when these factors varied. In fact, in recent years, due to the downturn of the global economy, the demographics of foreign populations within the district started to drop. Meanwhile, the once great distance between the Biyun neighborhoods and downtown Shanghai has gradually diminished through the years due to Shanghai's rapid expansion. Moreover, as the city as a whole gains more and more international momentum, the need to establish new foreign quarters or maintain existing ones has been highlighted. With all these factors in mind, we can assert with confidence that the Biyun neighborhood is now facing an identity crisis. It is crucial for Biyun to reinterpret its role within the city of Shanghai. Its original identity has lost its momentum both internally – as its foreign population decreases – and externally – as Shanghai is now asking Biyun to prepare to be incorporated into its continuous urban framework. (Fig.1)

It is clear that Biyun's new role will be a result of the negotiation between its desire to retain its uniqueness to a certain extent, and the whole city's urgent need for Biyun to break down its barriers to facilitate the expansion of outer urban fabrics. The former stands for the interests of the local, the buildings and blocks of Biyun, while the latter stands for the interests of the global, the city of Shanghai. The interaction between the two is intricate. On one hand, an agreement needs to be reached. On the other hand, though Biyun is seemingly subordinate to Shanghai, its inertia as a foreign enclave distinct from its surroundings is so strong that its demand for the preservation of its own identity needs to be respected.

As previously mentioned, the urban materialization of socioeconomic factors cannot be separated from architectural forms or typologies, as the socioeconomic matters, as symbols, needs signifiers to tell their stories. Apart from that, the different modes of existence, as well as the different understandings of the identity of a neighborhood is hardly detached from a formal realization of architectural elements. The architectural language is the medium through which all these various perceptions and materializations of ideas are conveyed and conducted. So, in this sense, the aforementioned issue of the conflict of identity in the neighborhoods of Biyun, is also articulated through the carrying-

当时的背景下，由于中国整体开放水平不高，上海的国际化程度也较弱，用有限的资源培育小型的高端外国人社区来吸引外企落地符合那个发展阶段的需要，换言之，当时碧云自身的兴趣与其所处城市的发展兴趣一致，也就具有一定的政策基础。而我们也不难想到，这一系列的社会经济因素一旦发生变化，碧云的发展态势也将受到显著的影响。实际上，由于近年来全球经济的走弱，上海浦东的外国人口首次下降。与此相对的是，上海自身作为一个城市整体，其国际影响力则越来越显著，营造独立的外国人社区的需求相对来说不再强烈。另外，经过十几年的大规模城市扩张，碧云与上海中心城区的距离也迅速拉近，其与周边环境的不连续性，存在阻碍城市有机扩展的可能。基于以上种种原因，我们可以认为碧云社区本身作为一个个体，正遭遇一种前所未有的自我认同危机。在这种情况下，由于碧云原先的身份已经丧失了存续的支撑因素，而且上海正在对碧云提出融入城市有机框架的要求，碧云重新解读建构自己在上海这个整体中的个体身份显得非常重要。（图1）

显然，碧云的这个新角色，将是其自身保持一定程度独特性的意愿，与打破自身屏障以顺应上海城市化进程这个对城市整体的意愿之间协商的结果。前者代表了社区这个个体的局部意愿，而后者则代表了城市这个整体的全局意愿。这两者之间的相互作用错综复杂。一方面，碧云社区作为上海的一个部分，最终必须与上海达成一个共识；而另一方面，尽管碧云从属于上海，但作为一个外国人飞地，其个性非常突出而与周围环境显著相异，具有非常强的个体惯性，以至于其一定程度上保持自身个性的意愿几乎无可避免地应该得到尊重。

前面提到，城市物化的社会经济因素不能从建筑形式或类型学的社会经济中被剥夺出来，因为社会经济作为一种符号需要能讲述自己的故事。除此之外，存在的不同模式以及对一个社区的身份的不同理解，难以从建筑元素的实现形式中分离出来，因为建筑语言是一种可以使各种观念和思想物化并被输送和传导的媒介。所以从这个意义上讲，上述提到的碧云社区的身份冲突问题也是通过其建筑形式的实现来表达的。考虑到这一点，在讨论实际的冲突协商过程之前，寻找强调冲突中双方的愿望的建筑语言是很有帮助的。

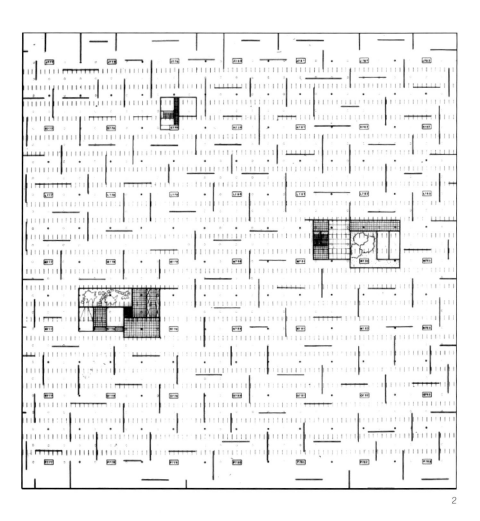

Fig.2 No-Stop City. Archizoom, 1968-1970. Plates from the project's publication in Domus, March 1971

图2　不停留城市，Archizoom，1968-1970年，刊于《Domus》杂志1971年三月期

out of its architectural operations. Considering this, before moving on to a discussion of the actual negotiation that is to take place, it is beneficial to look into the tools, or the architectural languages that represent the desire of the two parties within the conflict.

As for the city, if we solely start from the extreme condition where it wants to break down all the barriers that might halt its expansion and incorporate all the functional parts into itself rigorously, we would believe that it wants to render itself as a field free of boundary conditions and internal disruptions. This kind of scenario would inevitably remind us of the experimental description of the 'No-Stop City' as illustrated by Archizoom (Fig. 2).[1]

Within the 'No-Stop City' we find *"a homogeneous amalgam of all real data, where there is no longer any need for 'zoning'."*[2] Branzi also wrote of the 'No-Stop City' that:

The idea of an inexpressive, catatonic architecture, the outcome of the expansive forms of logic of the system and its class antagonisms, was the only modern architecture of interest to us: a liberating architecture, corresponding to mass democracy, devoid of demos and of cratos (of people and of power), and both centerless and imageless. A society freed from the rhetorical forms of humanitarian socialism and rhetorical progressivism: an architecture that gazed fearlessly at the logic of gray, unaesthetic, and de-dramatized industrialism....[3]

Even though it is arguable whether, in such a condition, it can be true that the greatest possible freedom occurred where integration was strongest, one thing is for sure: the autonomy of the parts is totally overwhelmed by the congruity of the city as a whole. There is no longer any essence of the block, for all the elements are broken down into objects no larger than the dimension of a block.

1 Archizoom, No-Stop City, 1968-1970. Plates from the project's publication in *Domus*, March 1971.

2 Aureli, Pier Vittorio. *The project of Autonomy: Politics and Architecture within and against Capitalism*, 75-76.

3 Branzi, Andrea. "Postscript", in *No-Stop City, Archizoom Associati*, 148-149.

3

Fig.3 The City of the Captive Globe. Rem Koolhaas 1972
Source: Koolhaas, Rem. *Delirious New York : A retroactive manifesto for Manhattan.* New York: Oxford University Press, 1978. 244

图3 囚禁中的城市，雷姆·库哈斯，1972
来源：雷姆·库哈斯，《癫狂的纽约：一部曼哈顿的回溯性宣言》，纽约：牛津大学出版社，1978年，244页

对于城市而言，我们假定一个极端条件——当其想打破所有可能停止其扩张的障碍，而且希望将所有功能部件有条不紊地吸纳进自身，我们便可认为其希望自己呈现为一个不受边界条件与内部奇点所约束的场。这种情景将不可避免地让我们想到建筑视窗（Archizoom）的不停留城市的实验（图2）[1]。

在'不停留城市'中，"作为一个汇集所有真实数据的各向同性的混合物，将不再有任何对于区划的需要"[2]。除此之外，布兰奇还写道：

"关于钝化无回应的建筑的概念，系统逻辑的扩张形式的结果及其阶级对立，是唯一对我们有意义的现代建筑：一种对应普遍民主，脱离权力与人民，无中心且无意象的解放的建筑。从人文主义修辞形式和修辞进步主义中解脱出来的社会：即无畏地凝视灰色逻辑，不受美学影响的，且去戏剧主义的产业主义建筑……"[3]

即便在这样的条件下对于最大的自由发生在整合最强的地方是不是正确仍有争议，有一件事是肯定的——部分的自主权完全不堪城市整体一致性的重负，因而区块的概念不复存在，因为所有的元素都被破坏成不超过一个区块尺寸的对象。

与同质化和缺乏自主性相反的是，对碧云这样富裕且对外界漠不关心的区块来说，其始终保有保持自我意识和作为一个超然实体的自主权的愿望。从这个角度，我们不难相信碧云仍然想要保持其以现有的围墙和大门为标志的门槛从而阻止外界的干涉。更重要的是，它可能还希望保留其内在的作为一个相对超然的、

1 建筑视窗，《不停留城市》，创作于1968-1970年，摘自《Domus》1971年三月版

2 皮埃尔·维托利奥·奥雷利，《自治项目：资本主义内与反资本主义的政治与建筑》，75-76页

3 安德里·布兰奇，《附言》，摘自《不停留城市》，建筑视窗事务所，148-149页

5

4

Fig.4 Moema, São Paulo
Source:http://vadebike.org/wp-content/
uploads/2011/11/DSC07272.jpg

Fig.5 Moema Building Morphology

图4 圣保罗Moema街景
来源：http://vadebike.org/wp-content/
uploads/2011/11/DSC07272.jpg

图5 Moema建筑形态

Contrary to this homogeneity and lack of autonomy, the ultimate need from the perspective of an established block as affluent and indifferent as Biyun would be preserving its self-consciousness and a certain degree of its autonomy as an aloof entity in its own right. In light of this, it is understandable that the Biyun blocks still want to maintain their thresholds, as signified by the existing fences and gates that block the exterior from intruding, but perhaps in some new forms. What is more, Biyun might also hope to retain its inherent nature as a relatively aloof neighborhood that is aware of its social status, even though it has to learn to coexist with its neighbors in a significantly less distant way, and even start to mingle with them. A block like this, with a high degree of self-conscious self-fulfillment, as contrasted to the subservient elements in the No-Stop City, would almost immediately bear some resemblance to the building blocks depicted as islands that constitute an archipelago in *City of Captive Globe* (Fig.3). They are a block within the archipelago, according to a plot reminiscent of Superstudio's narrative.

"'The City of the Captive Globe' warns against the totalizing, though ineffective, dimension of ideology when it is not in some way contained."[4]

Thus, we can derive from such a conceptual model of the city block an architectural element that bears features of a barrier, a container, or perhaps also a tower. As for a neighborhood that is as affluent and culturally distinct as Biyun, the self-consciousness of identity is also coupled with a sense of superiority, which may further render its self-perception as a sole island in the ocean of commonness. After looking into some conceptual models that, through postulating some extreme scenario of a pure form of existence of one single ideology, reflect the crucial elements of architectural operations that signify the ideology of the local and the global respectively, it is worth looking into some existing cases of building block typologies that can be regarded as products of conflicts similar in nature to the one between Biyun and the city of Shanghai.

First is the neighborhood of Moema in São Paulo, Brazil, which is a long established affluent neighborhood where the primary residents are American expatriates. From an overhead view, the lofty residential towers of Moema would give one an impression of a series of residential 'islands' that have established themselves as distinctive and self-dignified (Fig.4-6). However, beyond that, one would also easily notice by walking around its streets that on a lower level, the blocks are equipped with articulated street level configurations that merge with the larger urban context of São Paulo, an emerging global city that accommodates all classes and a wide range of cultures. Thus, the blocks within the neighborhood of Moema could be understood as lofty islands that stand on their substrate and are almost indistinguishable from their urban contexts. In other words, the residential towers have retreated and given away their grounds as a sign of respect for the city as a whole in exchange for the long-term maintenance of their own identities.

4 Koohaas, Remmet. "The City of the Captive Globe", in *OMA*, 331–33.

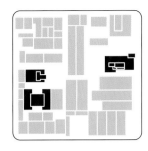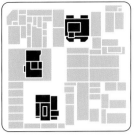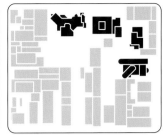
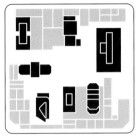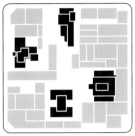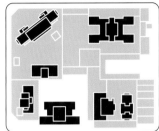
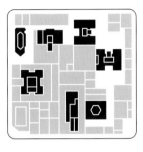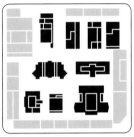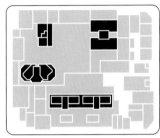

6

Fig.6 Moema Block Configuration
图6 Moema街区构成

对自己的社会地位有自我意识的居民区的性质,即使它需要学习如何在与其周边城市片区大大缩短的距离内共存甚至开始融合。像这种与"No-Stop City"中对整体绝对顺从的元素截然不同的有着高度自我意识与自我实现意愿的区块,与库哈斯所著的《囚禁中的城市》中描绘的"群岛中的岛屿"(图3)般的建筑区块有一定的相似之处。对于处在"群岛"中的区块,Superstudio的叙述中说到:

"《囚禁中的城市》认为,在位于某种意义上被承载的条件下,穷尽却缺乏效力的意识形态的量度是不可取的。"4

据此,我们可以从这样的一个城市街区的概念模型中推导出一个具有屏障、容器、甚至是一个塔楼的性质的建筑元素。对于碧云这样的具有独特文化的富裕街区,自我身份意识还会加上一种优越感,而这可能进一步将其自我认知呈现为"一片共性海洋中的独特岛屿"。在对一些通过将单一建筑意识形态推向极致,来反映与该种意识形态相关联的建筑形态进行分析之后,我们应当基于这些认识进一步从现实城市中寻找类似上海和碧云之间的冲突的案例,对城市的社会经济因素与具体的建筑形态进行解读。

我们的第一个案例是巴西圣保罗著名的美国人社区——Moema街区。Moema位于圣保罗金融中心圣保罗人大道(Avenida Paulista)南面,多数居民为生活在圣保罗的美国人,街区总体富裕,居住环境宜人便捷,与圣保罗大众所居住的一般城区形成鲜明对比。从一个整体的视角来看,Moema的为数众多的高层居民楼犹如一座座高耸的、刻意与周遭环境以示区分的"居民岛"(图4-6)。然而,在这个宏观的形态之下,当观察者作为行人置身于该街区之内,则又不难发现Moema

4 雷姆·库哈斯,《囚禁中的城市》,摘自《OMA》,331-333页

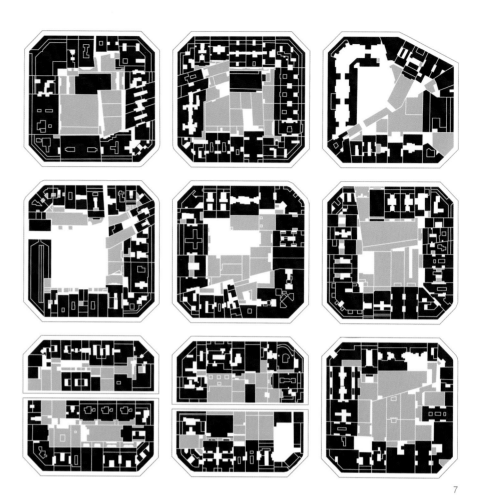

7

Fig.7 Barcelona Block Configuration

图7 Barcelona街区构成

From this point on, it's not hard to see that such a gesture of architectural operation actually incorporates, at once, a monolithic residential tower that is impenetrable by outsiders, and which is the container of the self-consciousness of the neighborhood, and an accommodating street level that somehow resembles a leakage, where signs of hospitality are revealed.

If we depict the building blocks of Moema as a top-down stratified architectural type that have introduced the urban ground into the blocks even as they maintain the self-consciousness of a different mode of living on a lifted ground, then the building blocks in Eixample in Barcelona as planned by Ildefons Cerdà in the 19th century would be better understood as a zoning operation that mainly happens in plan (Fig. 7-9). In the coding of a Cerdà-style block, there are both areas dedicated to the free-flow of the city and, at the same time, zones that are striving to fulfill the demand of a certain degree of privacy.[5] In the light of this, a typical Barcelona block would normally incorporate a series of urban programs ,such as retail, service, etc., that are usually the height of one or two stories, wrapping around the perimeter that is facing the street front. Meanwhile, both behind and above these relatively more urban programs, there are residential units that enjoy a higher degree of privacy. The relationship of the two types of programs, or the two zones that are trying to cater to two different types of needs: a local one and an urban one, are mediated by some gaps at approximately the height of the street level that take place in the perimeter of the block. These gaps can be understood as the product of the carving force of circulation, as well as a common demand from both inside and outside to cast the two parties into a harmonious state of symbiosis.

In both the cases above – the blocks of the American Expatriate Community in Moema and the blocks planned by Cerdà in Eixample, Barcelona – there is a common theme of retaining some architectural features that are pivotal for realizing the need of one party while comprising the integrity of both parties to facilitate the fusion of the two together.

5 Puig, Arturo Soria y (ed): *Cerdá: the five bases of the general theory of urbanization*, Electa, 1999.

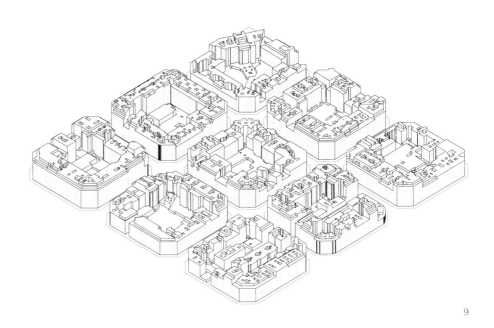

9

8

Fig.8 Inside a block in Eixample, Barcelona
Source:http://you.ctrip.com/travels/barcelona381/1752136.html

Fig.9 Barcelona Building Morphology

图8 巴塞罗那扩展区街区内部
来源：http://you.ctrip.com/travels/barcelona381/1752136.html

图9 巴塞罗那扩展区建筑形态

的大部分街区在靠近街道的层面上往往都表现出一种能够与圣保罗城市整体所无缝连接的建筑形态，以至于在行人的角度来看，从圣保罗的市中心到Moema街区内的步行体验很难察觉出显著的改变。这些临街界面的气质不同于其背后与上方的住宅单元，真正融入了与圣保罗作为一个包容不同阶级与文化背景的发展中国家国际大都市的特性。如此一来，Moema区块内的这些由高层居民楼与独特临街界面所组成的街区，即可被理解为一系列坐落于与城市整体相融合的基座上的高耸甚至有些清高的城市岛群。换句话说，居民楼实际上退让了，把直接接触街道与地面的空间放弃了，以示自己对城市整体的自由流通性和全局特质的尊重。这种牺牲实际上换取了其长期维持自身独特性的空间。从这一点上看，我们不难发现这样一种建筑的具体形式：同时结合了具有强烈自我意识的、对外人来说不可进入的孤岛式的高层居民塔楼，以及无时无刻不在释放善意，或者更准确地说，"表现善意"的亲民好客的街道层面。

如果我们把Moema的街区描述为一种垂直分层的建筑形态，一种既将城市整体的肌理引入底层，而又在上层继续营造保持区块本身个性的居住空间的混合机制，那么由塞达(Ildefons Cerdà)在19世纪所规划设计的巴塞罗那扩展区(Eixample)的典型街区，则可以被理解为一种可从平面图上出发的分区机制（图7-9）。在塞达式街区的编制系统里同时兼有适应于城市的自由流动性的区域，和更倾向于满足一定程度的私密性的部分。[5] 从这一点出发，一个典型的塞达式街区通常会包含一系列诸如零售、服务等的城市功能。这些功能区通常分布于街区首一二层的直接面向街道而把街区内部包裹起来的外围区域。与此同时，在这些满足城市功能需求的区域之后以及之上，则是具有更高私密性的居住单元。两种功能之间，或者说两种分别尝试满足街区局部与城市全局的需求的区块之间的关系，则被一些分布于区块外围的与街道层面大致等高的开口所调和。这些开口从物理作用的角度可以被理解为动线的切割作用的产物，而从抽象概念上看，则亦是来自街区内部与外部两股作用力之间共有的希望与对方共生而达到和谐状态的意愿的产物。

上述的两个案例——圣保罗Moema的美国人社区以及巴塞罗那扩展区的典型街区——都围绕着一个共同的主题，亦即冲突双方各自在一定程度保持对实现己方的需求起关键作用的建筑特性的同时，牺牲各自的完整性以调和双方矛盾，达到二者的融合。

5 阿图罗·索里亚·普依格编，《塞尔达：整体城市化理论的五个基本点》，Electa出版社，1999年

As for Biyun, based on the studies above, we should have no difficulty in terms of what architectural languages we would like to utilize to address its current issues. The only thing is that, before the application of those languages from external examples and theories, it is worth investigating the current zoning and architectural realizations within the neighborhood. In fact, there are two main features that are essential within the discussion of the conflict between the expansion of the city and the preservation of the neighborhood's identity. First is that the current typology within Biyun, to a large extent, lacks the mixing of programs within blocks. In other words, the zoning throughout the site is overtly planar, making some of the blocks – mainly the commercial and open land areas – very submissive to the city's desire to absorb them. Others, mainly the enclosed residential blocks, are almost obstinate in forming obstacles for such a process. In the meantime, another feature of the layout of the architecture within the neighborhood is that the program-specific blocks are also highly polarized; most of the commercial blocks are concentrated on the north rim, while everything behind them are rigid, indifferent blocks that are hard for the city to penetrate. As a consequence, the barriers formed by the enclosed blocks are not only holding the ground, but also imposing a pressure in terms of orientation toward the free flow of the urban context. Both of these problems can be related back to the previously discussed cases; from a planar point of view, the problem of polarity in the neighborhoods can be addressed through an inside-out redistribution of types specific to programs. With gaps carefully carved out in a similar manner to the typical blocks in Barcelona, tension driven by the two-way contrast of public and private sides (or in the case of Biyun, more aptly named local and global sides), intensified by the stress of a free urban flow, can be eased. Raising the ground level of the residential blocks by forcing them to give up the floor levels that have direct attachment to the ground, creating space to accommodate the expansion of the city into the blocks, can, at the same time, help preserve their aloofness and self-consciousness on the levels above, which is less disturbed by the urban context.

In conclusion, Biyun's current situation is a normal stage that many communities will experience or have already experienced, and the factors behind such a circumstance – the conflict operating both internally and externally – are far from being unique. The reconfiguration that incorporates the theories of *"No-Stop City,"* *"The City of Captive Globe,"* and the urban cases of Moema, São Paulo and Eixample, Barcelona, would be a simple move but have profound effect in addressing the current conflict.

对碧云社区而言，基于以上的研究，我们不难提取出对解决其现有问题而言比较重要的建筑语言。而在真正将这些从外部案例与理论模型中汲取的认识与经验应用于碧云社区本身之前，我们还应对碧云目前的分区与建筑形式进行调研。实际上，对当下面对着城市扩展的压力和维持自身特点的意愿之间的冲突的碧云社区而言，有两方面的性质非常值得讨论。第一是碧云目前的建筑形式。很大程度上来说，目前碧云的街区各自的功能都非常单一，缺乏混合功能的街区。换言之，整个碧云区块的分区高度平面化而且整块化，使得某些集中了商业功能和城市绿地的街区对上海城市扩张的意愿非常顺从，而那些集中了居住功能的封闭街区在面对这个进程时却又岿然不动。第二，除了街区功能单一的问题，从碧云整体来看，整个区域的街区形式与功能分布还存在两极化的问题。具体来说，前述的那些碧云内部的功能单一的街区，在碧云内部的分布呈现商业街区集中在北线，而所有的对城市扩张阻力巨大的居住功能区则都在这一条线的背后。这样一来，由封闭街区所组成的屏障不仅直接与街道层面接触，牢牢抓住这个区块的地基，另一方面，还影响到周边环境里的城市自由流动的朝向。尽管如此，这两个问题都可以联系到前面讨论过的案例。从平面图的角度看，碧云区块整体的极化问题可以通过一种类似塞达式街区的从内而外的功能重分布来回应。通过类似塞达式街区那样对区块外围开口的精心处理，街区的公共面与私密面双方的矛盾以及城市整体对各向同性的自由流通的需要能够得到缓解。另一方面，借鉴圣保罗的Moema街区的经验，从垂直分层的角度出发，将惯性较强、对私密性要求较高的居住单元与街道层面割裂，将之抬升到一定高度，而将空余出来的空间用于更顺应于城市整体扩张与自由流通的诸如商业、绿地一类的城市功能区，也在保持街区自身个性的同时，更好地与城市扩张的大背景相协调。

作为结论，碧云目前所面对的问题是一个许多街区将会面对或者已经经历过的普遍问题，而在这种情形背后的因素则因地而异。对碧云本身而言，基于"No-Stop City"，"The City of Captive Globe"的理论认识，以及圣保罗Moema街区和巴塞罗那扩展区的实际经验的内部重新配置，是一项相对简单的举措，但会对其目前面对的问题产生深刻的影响。

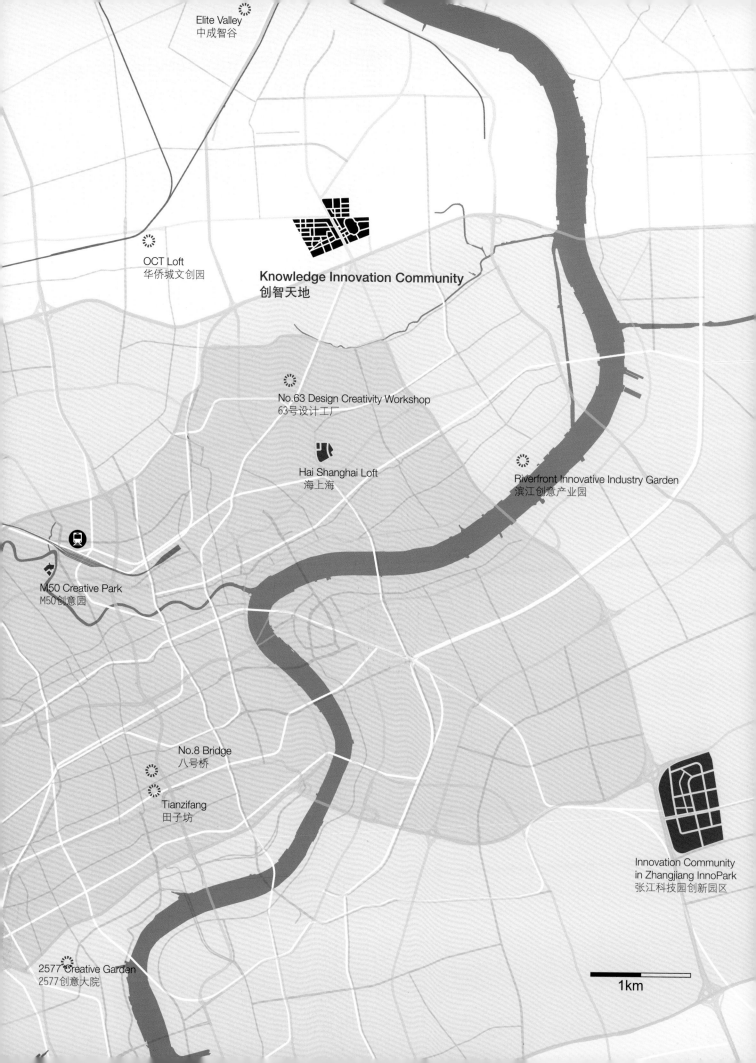

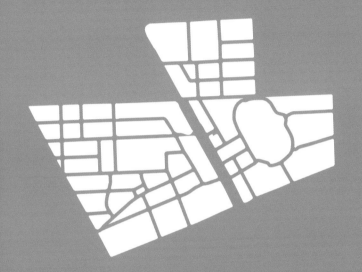

Innovative Motivation
创新驱动

Knowledge Innovation Community (KIC)
创智天地

Trasforming district from post-industrial to innovation
转化后工业地块为创新创意园区

Dingliang Yang, Fan Wang

Innovation District as A Glocalized Urban Strategy
Reading Knowledge and Innovation Community (KIC)

Fig.1 Knowledge and Innovation Community, Phase 2, University Avune Eastside Birdview
Source:http://casestudies.uli.org/knowledge-and-innovation-community-shanghai/

图1 创智天地社区二期，大学路东侧鸟瞰
来源：http://casestudies.uli.org/knowledge-and-innovation-community-shanghai/

KIC, the Knowledge and Innovation Community (Fig.1-2), is a large, mixed-use technology and innovation community located in Shanghai, China. It was developed by Shui On Land and planned and designed by Skidmore, Owings and Merrill (SOM) in collaboration with Shanghai Urban Planning, the Design Research Institute, and Tianhua, to foster a creative and entrepreneurial culture in Shanghai. It was built on a 121-acre (49 hectare) site in the Wujiaochang area of the Yangpu District, amid several of Shanghai's premier universities, including Fudan and Tojing Universities. It has a total gross floor area of 762,217 m^2, all of which creates a high-quality mixed-use urban environment combining residential, investment, retail, office and educational use, as well as incubator services.

It is probably very hard to comprehend KIC without knowing about its developer, Shui On Land, a Hong Kong-based real estate developer that has been playing an important role in the urban redevelopment of China; they are especially famous for their Xintiandi project, located in the heart of Shanghai. Shui On Land has been enthusiastic about, and specializes in, developing large-scale urban regeneration projects. It is known for the Xintiandi Project, a successful cultural center and major tourist attraction in downtown Shanghai, which preserves the traditional Shanghai lane house, a form of architecture known as "Shikumen." Hence it is not hard to understand why Shui On Land participated in and led the KIC project, trying to turn a declining industrial area into a new hub of growth.

In the early 2000s, China wanted to enhance economic growth through technology and innovation, and the city of Shanghai advocated for the same. Shanghai's Yangpu District was old and dilapidated, a declining industrial area far from the city's main business centers around the People's Square in Puxi and the Lujiazui Financial District in Pudong; it was not an obvious location for a leading creativity and innovation center. Beginning in 2002, Shui On Land, in partnership with the Shanghai Yangpu District Government, started to examine and study the issue and came up with the idea that the district could be developed into a technology and innovation center that would leverage the presence of the nearby universities and institutes, including the historic Jiangwan Stadium. KIC was imagined like a Silicon Valley or Paris Rive Gauche in the Yangpu District.

An important part of the development proposal that Shui On Land submitted to the Yangpu District Government, was to build a mixed-use urban neighborhood that would include residential and retail components, along with the offices and incubators, setting it apart from other business parks in China during that time. This plan later became the basis for KIC. The project got started in 2003, but it took over 12 years to build the infrastructure and develop the ecosystem for a real knowledge community, which is now working quite well as a vibrant mixed-use neighborhood home to leading technology firms like EMC2, Oracle, and IBM, as well as over 600 multi-national and domestic high-tech companies. The project was a finalist in the 2015 Global Awards for Excellence hosted by the Urban Land Institute; their description of the project is as follows: *"The*

杨丁亮，王凡

创新园区作为全球本土化的城市策略
阅读创智天地（知识与创新社区）

Fig.2 Knowledge and Innovation Community, Phase 1, Jiangwan Stadium Westside Birdview
Source:http://casestudies.uli.org/knowledge-and-innovation-community-shanghai/

图2 创智天地社区一期，江湾体育场西侧鸟瞰
来源：http://casestudies.uli.org/knowledge-and-innovation-community-shanghai/

创智天地（图1-2）是由瑞安地产开发，Skidmore, Owings & Merrill LLP（SOM）公司、上海市城市规划设计研究院与天华建筑联合设计的以科技与创新为主的大型综合社区。其目的在于为上海培养创造力和创业文化。创智天地坐落于杨浦区五角场的子中心一个121英亩（49公顷）的场地上，毗邻复旦大学与同济大学等上海一流学府。创智天地项目建筑总面积为762217平方米，其中包含了高品质的住宅、办公、零售、投资业以及教育与孵化器产业等服务。

在了解创智天地之前，在某种意义上很有必要了解一下它的开发商瑞安地产。瑞安地产是一家香港房地产开发商，它在中国的城市改造和建设中发挥了重要的作用，其中上海新天地项目让其名声大噪。正由于瑞安地产专注于大型城市更新项目才造就了如上海新天地等成功的城市设计案例。上海新天地保留了传统里弄建筑"石库门"并发现其价值，同时与商业开发相结合，将新天地变成了上海中心城区的主要旅游景点与商业区。因此，并不难理解瑞安地产能成功领导创智天地项目从一个旧工业区转化为新的增长中心的能力。

Fig.3 1927 Greater Shanghai Plan of Wujiaochang, Yangpu District
Fig.4 1935 the Surrounding Area of Jiangwan Stadium

图3 1927年大上海规划，杨浦区五角场鸟瞰
图4 1935年江湾体育场周边地区鸟瞰

successful transition from a manufacturing powerhouse to a center for the modern creative service economy serves as a model in microcosm for China."[1]

According to the committee of the ULI Global Awards, the key of the KIC project is the central idea of combining the resources of the universities and the start-ups within KIC. The technology companies work together with the universities and institutes to build a real community where young people with new ideas and innovations can go to work, play, and live. As a mixed-use innovation cluster, KIC was a more complex project to develop than a typical single-use development project. Thus, both the developer and the planner and designers were challenged to build the "ecosystem" desired by the Shanghai Yangpu Government and Shui On Land.

The case study here is endeavoring to research this specific case and, by analyzing KIC, to understand the paradigm of transforming a previously derelict post-industrial land into a mixed-use innovation district. The study of KIC is conducted in five aspects, including all the major layers of the project: the urban context, the concept and the masterplan, the grids and blocks, the open space, and the architecture/buildings. The research methodology of the analysis is mainly a comparative reading, both putting KIC into dialogue with similar global cases and comparing it with other redevelopment projects in Shanghai.

[1] Urban Land Institute. *ULI Case Studies: Knowledge And Innovation Community*. Washington, D.C. : Urban Land Institute, 2015.

在21世纪初，中国政府想通过推动技术创新来提升经济增长，这一想法也得到了上海市的大力支持。当时的杨浦区是一个老旧的废弃工业区，地理位置远离浦西的人民广场和浦东的陆家嘴等主要的经济中心。所以，当时的杨浦区并不是一个特别理想的区域，并不适合作为前沿创新与创造力中心的基地。但是从2002年开始，瑞安地产与上海杨浦区政府合作研究了创智天地项目的可行性，提出了依托杨浦区历史悠久的江湾体育场和周边一流大学等教育资源的优势，提出了将创智天地打造成上海杨浦区的硅谷和巴黎左岸的构想。

创智天地项目不同于上海其他城市更新改造项目的一个重要方面，在于瑞安地产向杨浦区政府提出的发展建议，是将其建设成一个城市大型综合型社区。此项目将住宅、商业、零售、办公与产业孵化器等不同功能融合形成一个综合产业园区，与当时上海其他的办公园区理念颇为不同。这一融合随后形成了创智天地项目的基石。从2003伊始后的12年间，该项目建造了较为完善的基础设施和生态系统，真正的知识创新产业园区逐渐形成。如今这一系统运行良好且极具活力，吸引了如EMC2、IBM、甲骨文等世界著名科技公司以及超过600多家国内外高科技公司。该项目获得了城市土地学会颁发的"2015年全球卓越奖"并被描述为："*从制造业成功转型为现代创意中心，创智天地的成功可做为中国经济转型的一个典型的微型案例*"。[1]

ULI全球奖委员会将该项目的成功归结于合理整合资源并且依托大学的人才和初创企业的活力，以及与科研机构的积极合作，建立起了一个真正让年轻人可以分享经验、想法，一起工作、生活的社区。作为一个大型综合性的创新产业集群，创智天地项目比一般单一产业结构的商业开发更为复杂。因此，此项目挑战了开发商、规划师与设计师的协作能力，如何创建一个"生态的系统"。最后的结果也正如杨浦区政府所愿，项目的各个参与方十分成功地塑造了一个有活力的生态系统。

此次案例分析旨在详细分析和研究创智天地项目是如何将一个废旧工业区改造为综合成功的创新社区的。同时详细地解读了该项目的五个主要方面：城市背景、概念和总体规划、网格与体块、开放空间和建筑等。通过与全球同类项目的比较与对话，深入了解该项目的脉络。

[1] 城市土地学会，《城市土地学会案例分析：知识和创新社区》，华盛顿特区：城市土地学会，2015年

CONTEXTS AND MASTERPLAN - Advantages and Challenges

The site of KIC is located in Wujiaochang, in the Yangpu District of Shanghai, where there is a history of urban planning and city-making; however, until the 2000s the area was still underdeveloped and even declining. Before World War II, Shanghai was a "treaty port," with much of the wealthy downtown area under foreign control and not subject to Chinese law. Yangpu District was a large industrial center on the periphery and home to thousands of migrant workers crowded into shantytowns. Its political status under Chinese jurisdiction made it attractive for development to the Nationalist government under Chiang Kai-shek. The 1927 Greater Shanghai plan (Fig.3), led by Architect Dong Dayou, called for making Wujiaochang ("five corner plaza"), then a regional marketplace, into the new Shanghai government headquarters and administration center. From the historic plan, we can easily find that the current KIC site was being considered as the site for city hall. The redevelopment plan was abandoned following the Japanese invasion in 1939, but by then, several new government buildings had been constructed. One of them, the Shanghai Municipal Stadium, was built within the boundaries of what is now KIC and has been restored as Jiangwan Stadium (Fig.4), a central part of the project's east side.

Following the war, Yangpu's industrial strength continued to grow, and by 1990 the district included 30 percent of the city's factories. But similar to the fate of industrial cities around the world—as well as the more central districts of Shanghai—starting in the 1990s, Yangpu found itself with a faltering industrial base and many obsolete and polluted industrial sites. It was struggling to create a vision for its future. However, Yangpu did have one powerful asset: several higher-education institutions were located close to the site, including Fudan University, Tongji University, and Shanghai University of Finance and Economics (SUFE), among others. Although these institutions combined have about 130,000 students and faculty, the district was benefiting little from their presence.

In 2003, Shanghai's leaders designated Wujiaochang as one of five major city subcenters for retail and leisure; it would serve as a hub for four districts and 2 million people in the northern part of the city. As part of the redevelopment, the government commissioned a striking new oval structure to enclose the elevated highway. Brightly lit at night, the structure, dubbed "the egg" by local residents, immediately became a regional landmark. Around the same time, the Shanghai and Yangpu District governments identified the need for a "Science and Education City," based on the development policy of "build a better city through knowledge and innovation." They endeavored to find a way to take advantage of Yangpu's strong educational resources.

The 49-hectare (121 acre) site that would become KIC occupied a strategic location between the Wujiaochang retail and entertainment center, the Fudan University and Shanghai University of Finance and Economics campuses, the Shanghai Second

项目背景与规划 – 优势与挑战

创智天地选址位于杨浦区的五角场，这是一个具有悠久的城市规划历史和城市建设经验的地区。但是到本世纪初，这里依然还是一片相对较为落后甚至已经开始衰败的工业区。在二战以前，上海市是一个通商口岸，富裕的城市中心区受到外国法律的保护且并不受中国法律的管制。旧城中心被外国人控制，所以位于外围的杨浦区成了位于城市周边的大型工业中心，这里聚集着成千上万的各地移民和工人以及他们的棚户住宅。杨浦区的地位（在中国的统治之下）得到了蒋介石国民党政府的重视与发展。1927年由建筑师董大酉主持的大上海计划（图3），本将五角场从一个区域市场转变为新上海的政府总部和行政中心。在大上海计划中，创智天地的场地正处于其中的市政府规划用地。但由于日本入侵，大上海计划在1939后被搁置。尽管如此，多个新政府大楼已经建成。其中的上海市体育馆（现江湾体育馆）正是位于创智天地项目东区（一期）的核心部分（图4）。

战后，杨浦区的工业实力持续增长，在90年代初，杨浦区的工厂总量占到了整个上海的30%。但与世界其他工业区和上海的中心区域的发展类似，90年代后期，杨浦区的工业逐渐衰落，产生了许多陈旧和污染的工业用地。区政府也在挣扎中试图寻找新的出路。然而，杨浦区具有高品质的教育资源，多个大学与研究院均位于此，诸如复旦大学、同济大学、上海财经大学等等。虽然这些大学和研究院给城区带来了13万的学生与教师，但是杨浦区并没有很好地利用这些高校以及它们所带来的资源。

2003年，上海的领导者将五角场指定为五个主要城市次中心之一。五角场区域以零售与休闲为主，主要服务城市北部四个区的两百多万人口。为了标志该区域的崛起，也作为城市改建和更新的一部分，五角场修建起了椭圆形结构的高速公路。这一夜晚灯火通明的城市景观很快就成了上海区域性的地标，并被当地居民戏称之为"大蛋"。杨浦区政府发现了建立"科教城"的需求，并根据规划纲领"建立知识与创新城市"，这一举动也是为了更合理地利用优秀的教育资源。

创智天地49公顷（121英亩）的区域位置正处于五角场的商业与娱乐的中心，毗邻上海财经大学、上海第二医科大学、江湾体育场，也靠近南部的同济大学。创智天地初期占据了一些小型工厂和工人住宅，并用体育场的周边地区作为一个公共汽车站。淞沪路作为一条主要交通要道将该场地一分为二，同时，大学路作为创智天地二期（西侧）的中轴，引领着一片非常活跃的街区。

Medical University, and the Jiangwan Stadium. The site was also only a short distance north of Tongji University. When the project began, the site encompassed a mix of small factories and worker housing; likewise, the stadium and surrounding area was used as a bus depot. The site was bisected by Songhu Road, a major artery that crosses the elevated Middle Ring Road expressway at Wujiaochang. Additionally, the University Road (Da Xue Road) on the west side of the site was a very active street that served as the backbone of the second phase of KIC.

One thing deserving mention here is that the site of KIC was planned to be served by a subway line, which eventually became Metro Line 10, constructed beginning in 2005. In addition to the benefits from the proximity of the surrounding universities, the planned subway was a huge advantage for the site in attracting visitors and new inhabitants. Inspired by the synergy of academia, industry, and culture driving California's Silicon Valley, the masterplan of KIC was developed to contribute to the fundamental principle of building integrated communities where people can work, live, learn, and play. The masterplan also focused on adapting historic structures for new uses and integrating them into new construction projects, as well as catalyzing the transformation of Shanghai's Yang Pu district from a dilapidated industrial zone to one with a thriving, knowledge-based economy.

"Through design, programming, and ongoing activities, KIC brings together students, researchers, entrepreneurs, and residents in a "three-zone linkage" of urban office, retail, and mixed-use community."[2] In order to fulfill the major target of creating a new innovation district serving as a newly active central zone for Shanghai, the masterplan of KIC centers on several key sub-objectives. The first is to develop an open, mixed-use plan; the second aim is to create a pedestrian retail street that would connect Fudan University , the restored Jiangwan Stadium, and a nearby subway station; and the third is to create a high-quality, active, sustainable environment that can attract talent and firms.

Before entering into the details of KIC's design strategies based on the fundamental principle and its three supporting principles, it is important to acknowledge the limitations of the site, which could actually strengthen the project by revealing how the design proposal transformed the shortcomings and limitations into advantages.

The first two limitations are the historic Jiangwan Stadium and a major artery road, which presented the project-makers with their main design challenges. The historic stadium sets the height limit for the whole development, limiting the building envelope of the project to no higher than 50 meters, which is very low in Shanghai. Second, the Songhu Lu artery road separates the entire site into two pieces and makes the project confront the issue of disconnection.

[2] SOM, *Designbook for Knowledge And Innovation Community*, 2003.

KIC Masterplan
创智天地总平面图

■ 创智天地一期 KIC Development Phase 1
■ 创智天地二期 KIC Development Phase 2
  主要道路 Main Streets
  主要水系 Main Waterbody
  开放空间 Open Space
  地铁站点 Subway Station
…… 地铁线路 Subway Lines

值得注意的是，地铁10号线在建成之后也服务于创智天地。此条线路已于2005开工修建。因此，除去拥有周边高校环绕的地理优势之外，创智天地还有地铁这一巨大的城市发展优势，因为地铁会为它带来许许多多的游客以及大量的城市居民。

借鉴加州硅谷利用学术和产业的相互作用、协同发展的成功经验，创智天地的开发和设计规划遵循了活动整体化的原则，将人们的工作、生活、学习、娱乐融为一体；并将旧建筑改造更新用于满足新的用途和功能要求，与新建筑结合到一起（共同组成丰富的城市空间），继而大力推动了杨浦区由原先落后的工业区向新经济形式下的繁荣城区的转型和发展。

"通过设计、功能规划和各种活动，创智天地将学生、科技人员、企业家和居民带入一个办公、商业和混合住宅'三区联动'的创新社区。"[2] 为实现创建上海的新型创新社区的目标，创智天地规划的几个关键要点是开发一个开放、混合的街区平面；其次是创建活跃的步行商业街连接复旦大学和地铁站并激活江湾体育场；最后是建立一个高品质、活跃和可持续的环境吸引企业与人才。

在讨论创智天地的设计理念和策略之前，我们必须要先明确场地的各种限制和它们给项目带来的困难和挑战。然后反过来，了解设计手法是如何将这些限制转化为项目优势的。

首先是历史悠久的江湾体育场和淞沪路分别带给创智天地的开发者和设计师的两个挑战。由于历史建筑的存在，整个项目的建筑限高被设定在了50米，远低于上海新建社区的平均高度。其次，由于淞沪路（很宽的车行快速路）的存在，将创智天地整个基地一分为二，带来了规划上可能出现的不连续性。

为了克服这两个挑战，规划师与设计师从两个层面提出了解决办法：首先是最大限度建立开放的公共空间，利用视觉上与实际上的连续性将两个场地连接起来，同时建立一个公共地下通道连接东西两区，东区为一期西区为二期；其次是有意地加强被分割的两个场地各自的特点，一边是历史悠久的江湾体育场与大型公园建立的开放空间，另一边是密集且活跃的步行街网络连接起小至中型的混合功能街区，创造舒适的生活环境。

[2] SOM，《创智天地：知识和创新社区设计文本》，2003

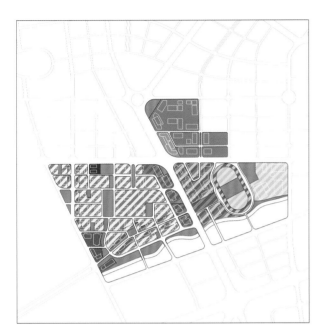

Urban Programs
城市建筑功能

- 商业 Commericial
- 办公 Office
- 酒店 Hotel
- 托幼 Daycare
- 绿地 Open Space
- 商业/办公 Commericial/Office
- 办公/居住 Office/Residential
- 商业/文娱 Commercial/Recreational
- 文娱/体育 Recreational/Sports
- 商业/绿地 Commercial/Open Space

Building Heights
建筑高度控制

- 历史保护建筑 Preserved Buildings
- 100米高度区 100m MAX
- 50米高度区 50m MAX
- 24米高度区 24m MAX
- 保护建筑高度控制区 12-15m MAX
- 保护建筑核心保护区 Preserved Core Area
- 开放空间 Open Space

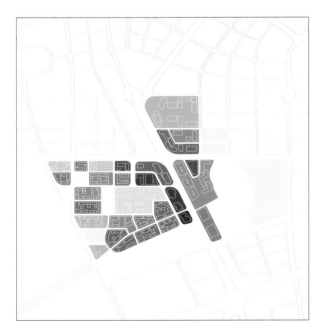

Floor Area Ratio
建筑容积率

4.6

0.3

Open Space Ratio
开放空间比率

94%

31%

In order to overcome these two challenges, the design tactics the planners and designers came up with are twofold: first is to maximize the functionality of the open and public spaces, which are taken advantage of to visually and physically connect the two sites and set the two sites into two phases connected by public underground corridors and walkways; second is to deliberately strengthen the differing characteristics of the two sites: on one side the site has a historic stadium with a park system as the main open space network and on the other side designers planned a dense and walkable street network that accommodates compact urban fabrics, intermediate-scale lots, mixed-use blocks and comfortable living areas.

One further challenge caused by the human-scale urban block is the issue of parking, which is very different from the maxi-block that has a large capacity for accommodating surface parking, while also structurally convenient for underground parking. In order to provide enough parking space to support this new district, KIC has to get parking built under small plots, then creatively link these parking lots together to form a large-scale one.

INNOVATION DISTRICT - Global Concept vs Localized Practice

Interest in innovation and culture in relation to urban development has taken off in a big way during the last two decades. Terms such as innovative district, innovative community, and innovative city, can be found in the latest urban development plans across the world. A new discourse has arisen that is shared by policy developers, advisers, researchers and designers, although the use of the terms innovative industry and innovative class is not always unambiguous. KIC also embraces this globally popular paradigm, aiming to use innovative industry to replace the previous manufacturing industry, stimulating further urban development in Shanghai. During the development of the KIC project (Fig.5), two parallel cases, or so-called references, have been treated as examples or exemplary comparisons: Paris Rive Gauche (Fig.6) and Silicon Valley in the Bay Area (Fig.7). It is even more clear here that the innovation district has been regarded as the motor of economic growth worldwide and has been imported to Shanghai at KIC. There appears a very interesting dialogue between the global concept, which is the innovation district, and the localization of the concept in a different way.

In his article *Creativity, Innovation, and Urban Development*, Paul Rutten criticizes the overuse of the notion of the Innovation District, pointing out that *"the ambition of cities to grow into creative cities runs the risk of turning into the opposite. A local or regional strategy implemented using a standardized strategic toolkit is as predictable as Rocky IV."*[3]

3 Rutten, Paul. "Creativity, innovation, and urban development." *Creativity and the City. How the Creative Economy is Changing the City (2005):* 66-79.

5   6   7

Fig.5 KIC in Shanghai, Before (bottom) / After (top)
Fig.6 Paris Rive Gauche, Before (bottom) / After (top)
Fig.7 Facebook Campus in Silicon Valley, Before (bottom) / After (top)
Source:
http://www.kic.net.cn/zh/learn
http://www.reinventer.paris/fr/sites/1240-paris-rive-gauche-13e.html
http://america.pink/paris-rive-gauche_3412579.html
https://newsroom.fb.com/media-gallery/menlo-park-headquarters/facebook-hq-aerial-view/
http://www.skyscrapercity.com/showthread.php?p=129786242

图5 上海创智天地，开发前（下）后（上）
图6 巴黎塞纳河左岸，开发前（下）后（上）
图7 硅谷Facebook总部，开发前（下）后（上）
来源：
http://www.kic.net.cn/zh/learn
http://www.reinventer.paris/fr/sites/1240-paris-rive-gauche-13e.html
http://america.pink/paris-rive-gauche_3412579.html
https://newsroom.fb.com/media-gallery/menlo-park-headquarters/facebook-hq-aerial-view/
http://www.skyscrapercity.com/showthread.php?p=129786242

还有一个挑战是步行尺度的街区所带来的停车问题。不同于大型街区有充足地表和地下来满足大量的停车位，小型街区的用地较为限制。为解决这一问题，创智天地必须将小型的停车场有机地连接起来才能具备足够的停车面积。

创新区域 – 国际化与本土化

过去的两个世纪，创新与文化对城市的发展起到了重要的作用。诸如创意小区、创意社区、创意城市等在世界各地都相当常见。虽然关于此术语的定义并不明确，但新的术语已引起开发商、顾问、设计与研究者等各方的关注。创智天地也追随了这一流行的趋势，希望通过创新产业代替老旧工业并刺激整个地区以至整个上海的城市发展。创智天地（图5）在构想和开发的过程中，有两个经常被提及的参考案例：巴黎左岸（图6）和加州硅谷（图7）。可见国际上火热的概念：把创新园区作为（新时代）城市发展中的重要元素和动力，被引入中国并通过创智天地在上海落地。这里有一个很有意思的话题就是"创新园区"作为国际化的概念和作为本土化的实践在理论和实践层面的辩证对立。

保罗·吕滕在他的文章《创意、创新与城市发展》中批判性地指出了"创新"这一词汇被过度运用："一个城市想成长为创新城市的同时很容易成为创新的反面。不同区域运用相同战略来达到创新城市的目的，其结果往往并非真正的创新。"[3]

根据吕滕的论断，我们可以也需要以此来审视创智天地，来判断在上海五角场这一

[3] 保罗·吕滕，《创意，创新与城市发展》，《创意与城市：创意经济是如何转变城市的》(2005)：66-79 页。

Fig.8 Paris Rive Gauche, Block Morphology

图8 巴黎塞纳河左岸，街区形态

Following Rutten's critique, we need to question whether establishing KIC, a pioneering innovation district in Wujiaochang, Shanghai is the right choice or not. As Richard Florida's book, "The Rise of the Creative Class," states, *"only a few cities were able to build up a distinctive profile like that of the crucible of the digital revolution, but that is what is necessary if you want to be competitive in the struggle to win talents and investors; the same goes for the development of the cultural and creative potential district of cities and regions; the starting point has to be the development of the unique and specific potential that a city or region has to offer."*[4] In other words, Florida emphasizes that the concept of making an innovative development can only be developed on the basis of a city's potential and proximity to talent and resources, which is why it is only to a limited extent something that can be created at all, and why it certainly cannot be brutally copied.

Paris Rive Gauche accommodates the vast François-Mitterrand Library and the Paris Diderot University along the river, while Silicon Valley sits aside Stanford University, a giant in the academic and research realm. Their locations next to universities and institutes, bolsters the common understanding of Florida's claim that the base of making an innovation district is the proximity to these potential talents and knowledge providers offered by a city. To this extent, KIC has an exceptionally sound base for making an innovation district because it is located in one of the subcenters of Shanghai, adjacent to universities and colleges, among which are some of the top think-tanks in China.

After confirming that the preliminary approach for KIC has certain similarities with Rive Gauche and Silicon Valley, the next step is to reveal the localization of the innovation concept in Shanghai and see its successes and failures by putting KIC in comparative dialogue with its two references, in terms of the varied physical urban environments they created and different levels of the social cohesion that they provided.

Block Strategies - The same concept of the innovation district has been differently demarcated and practiced in the three different contexts of Rive Gauche, Silicon Valley and KIC. Built from 1995 onwards on former railyards, warehouses, and industrial lands, Paris Rive Gauche is a 130-hectare innovation district in the thirteenth arrondissement of Paris, consisting of ten hectares of green space, bordered by the Seine, the railway tracks of Gare d'Austerlitz and the Périphérique, which can be read as taking advantage of the existing urban infrastructure (Fig.8) In order to make a dense, compact, mixed-use, mixed-income, pedestrian- and cycle-friendly, and public transport-rich set of new developments in hybrid urban form, the project is designed in three districts along the Seine: Austerlitz, Tolbiac and Massena, creating different urban morphologies. Austerlitz is the more infrastructure-related block, and Tolbiac is in the organization of a superblock dominated by the national library, while Massena is famous for the invention of the open

4 Florida, Richard. *The Rise of the Creative Class*. New York: Basic books, 2004.

地理位置设置创新创意园区是否是一个正确的决定，（能塑造一个真的城市创新片区），还是一个（如吕腾所说的）创新的反面。正如理查德·佛罗里达在《创新阶层的崛起》一书中所述："*只有少数城市可以建立起一个如数字革命般鲜明的创新中心，而这正是能吸引人才和投资的必要条件，也是发展城市文化和创新潜力区的必要条件。在创建一个创新园区之前，必须了解该区域是否能提供其城市本身的独特性和潜力，从而让这一目的得以实现。*"[4] 换句话说，佛罗里达的论述强调了一个创新园区的概念得以落实的先决条件有二，一是城市本身具有发展潜力这一基础，二是该地块本身需要毗邻人才和资源。这也解释了为什么只有少数特定的区域能真正建成创新区，而不是简单地沿袭其他城市的成功案例就能造就一个创新区域。

从这个意义上说，巴黎左岸的弗朗索瓦·密特朗图书馆和巴黎狄德罗大学，以及硅谷旁的学术研究殿堂斯坦福大学是创造创新园区的必要条件。前者是在基地内引入了新的知识机构，后者是在现有的大学边发展，但是它们都如佛罗里达所说，均地处潜力的城市，充分发挥了当地自身条件的优势，依托人才储备充足的大学和研究机构等建立创新园区。从这个角度来说，创智天地也具备这样的潜力和极好的基础：地处上海的一个副中心，而且其周边拥有包括了中国最顶级的智库和研究院的大学高校和优质的高等人才。

在肯定了创智天地的构想的可行性和发现了其与巴黎左岸与硅谷具有高度相似性（满足佛罗里达的理论条件）之后，我们下一步来分析创智天地是否是一个成功的项目。我们的研究方式是通过比较分析的方法，把创智天地和巴黎左岸及硅谷就它们塑造的城市空间和创造的社会效益进行对比，从而来探讨"创新园区"这一个国际化的概念在上海是否得到了很好的落实和开展。

街区设计 – 创新园区的概念在巴黎左岸、硅谷与创智天地三个项目中的本土化过程中有着不同的基地环境和由此带来的不同的理解和实施方式。自1995年起，建于原火车站、仓库和工业区的130公顷的巴黎左岸（图8）毗邻塞纳河且坐拥10公顷的绿地。原有奥斯德利兹火车站的铁路轨道和环城大道成为巴黎左岸的交通设施，成为项目的一个有利优势。为了形成紧凑、布局合理、可步行和骑脚踏车的混合型人性化社区，该项目沿塞纳河被划分为三个区域：奥斯德利兹、托尔比

4 　理查德·佛罗里达，《创意阶层的崛起》，纽约：基本书籍出版社，2004年

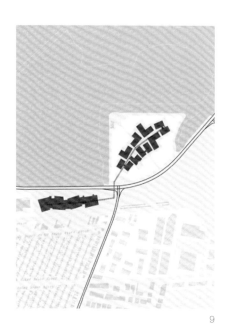

Fig.9 Facebook Campus in Silicon Valley, Building Footprints

图9 硅谷Facebook总部园区，建筑空间形态

mini block. Compared with Rive Gauche, KIC has almost the same size of the total area, and, probably learning from the experience of Rive Gauche, is also divided into two different phases: the eastside and the westside. Again, similarly, there are two versions of the block structure; the east is a large superblock while the west is an open, smaller-sized block. This is different from the masterplan and urban design approach that has been adopted in Silicon Valley, the most innovation-based place in the world, because of the huge difference in terms of scale. Silicon Valley in reality is an innovation city rather than a district, not even mentioning the term community. Instead of treating different phases with different block ideas, the overall strategy and fundamental principle of an innovation city in Silicon Valley is to use the campus as a basic formation unit to organize urban space. Every unit campus is a superblock compound, the most recent cases being the new 56-acre Facebook headquarters designed by Frank Gehry on mostly reclaimed land and the 175-acre Apple campus on the former Hewlett Packard (HP) production campus designed by Norman Foster. It makes sense to set this kind of unit campus as the comparison for KIC.

The different settings for the block strategies generate different outcomes. For Rive Gauche, the 380 x 200 meter superblock has been used to serve the national library, designed by Dominique Perrault, and successfully balances the public need for a large open space and the rigidity of monumental architectural layout; meanwhile the 40 x 50 meter open mini-block tactic created by Christian de Portzamparc has been used in the more urban-life oriented district, Massena, to help support the mixed-used community. The flexibility and porosity of the open block spatially has created an innovative neighborhood that attracts diverse and promising start-ups to set up base and younger generations to live here. By differentiating the way of designing differently-programmed blocks, Paris Rive Gauche created a relatively robust innovation district.

When looking at the Facebook or Apple campus compounds (Fig.9), it is clear the design is trying to make a vibrant internalized community rather than a very open district such as Rive Gauche, though the superblock is still the fundamental unit. There's a tendency that the building formation of the superblock in Silicon Valley is more and more often becoming a singular complex building. This type of innovation community is successful during the day because it is like a company-town, and the singular form is useful for working efficiency. Nevertheless, the night life is certainly not incorporated into the design of this type of superblock.

KIC is a duality that includes the 100 x 55m open block fabric, as well as one 400 x 150 meter superblock. The Phase Two open blocks on the westside of the project are very well accepted and favored by the public. In a similar case as the Paris Rive Gauche project, the openness and the various programs integrated into the district make the west part of KIC a real vivid mixed-use neighborhood. But on the other hand, the superblock is confronted by certain limitations. Unlike the Apple or Facebook campuses, which are designed as one single entity, and also unlike Rive Gauche's superblock, which is

克和马塞纳。三个地区有着不同的城市形态：奥斯德利兹以基础设施相关的街区为主；托尔比克形成了以图书馆为主导的超大尺度的超级街区；马塞纳以创新园区为主形成了小尺度的开放街区。与巴黎左岸相比，创智天地拥有几乎相同尺度的场地。汲取了巴黎左岸的成功经验，创智天地被分割为两个各具特点的区域：东区与西区。这两个区域采取了不同的街区规划尺度：东部以大尺度的超级街区为主，西部以小尺度的开放空间为主。世界上最具创新性的美国加州硅谷，因为尺度规模上非常大，它更像是一个城市而不是一个区域，所以它采用了不同的规划和设计方式。不像巴黎左岸那样把不同的分区用不同的街区设计策略，硅谷更偏向于校园式规划模式，用一个个校园作为空间的基本单元来组织城市空间。每一个单元都是一个超大街区综合体，例如弗兰克·盖里设计的占地56英亩的Facebook的总部和诺曼·福斯特设计的占地175英亩位于原先惠普的生产园区之上的苹果总部。因此就比较分析而言，硅谷中的每一个单元的园区更适合作为创智天地的比较对象。

不同的街区尺度会产生截然不同的结果。380米长200米宽的巴黎左岸超级街区很好地满足了人们对开放空间的需求和多米尼克·佩罗设计的法国国家图书馆对于空间的要求；而克里斯蒂安·德·保赞巴克在马塞纳设计的50米长40米宽的微小尺度开放街区创造出具有生活气息的城市空间。通透的小尺度街区设计更具灵活性，创造出的有活力和创造力的街区吸引了很多初创公司和年轻人工作和居住于此。通过不同尺度的街区规划来适应不同城市功能的要求，巴黎左岸创造了一个活跃的创新园区。

当我们着眼于Facebook或者是苹果的总部园区（图9），可以很清晰地发现，它们的规划设计意在通过划归超大街区进而在其中塑造一种内向型的社区而非像巴黎左岸那样的开放街区。而且在硅谷也有一种趋势，就是超大街区正在变化成超大建筑，也就是一个超大的城市综合体作为一个多功能的复合园区覆盖整个街区。这一形态的创新园区在白天是十分成功的，它的设计更多的是从公司的工作运行效率出发，就好像一个工厂城市，但是显然该类设计并没有考虑夜晚的生活，于是下班后这样的街区便会失去活力。

创智天地的设计兼具100米长55米宽的小尺度街区与400米长150米宽的大尺度街区。西区（二期）的小尺度街区被市民大众广泛接受和喜爱，也像巴黎左岸一样，空间的开放性和城市功能的多样性塑造了一个真正活跃的混合城市街区。而在另一边，东区的大尺度街区，遇到了一定程度的问题。不同于硅谷的Facebook

a pure public place, the KIC superblock tries to combine both characteristics to be a semi-public, semi-collective and semi-private area, as well as accommodating a cluster of companies in the same superblock. This kind of blurriness to a certain extent impedes the project from becoming a very good urban place. As Michael Storper has indicated, *"producers in advanced urban economies will require a higher tempo of innovation than the tempo of imitation to hold on to their competitive position. This enables a higher standard of living, too."* [5] The innovation district should either be a pioneer district that aims for an extreme mixture in both space and program, or it should stand to the side of mono-spatial organization, creating efficiency and functionality in highly-standardized space.

Sustainable Infrastructure - Another very important issue here is "transit". It is crystal clear that Paris Rive Gauche, the Facebook and Apple headquarters, and KIC are all in different locations in relation to the city center. Listing them by the distance of the site to the city center, Rive Gauche is in the center of Paris, KIC is close to the subcenter of the city, while Silicon Valley has a certain distance from San Jose and is even further away from the city of San Francisco. Transit mobility is the element that makes the new discourse of the innovation district seem to function regardless of proximity to traditional city centers. Rive Gauche is served by the Métro and RER stations Bibliothèque François Mitterrand and Quai de la Gare, so more than 30,000 people can conveniently work, live and stay in the community. The entire Silicon Valley is very well-served by the highway and dozens of small airports, as well as the commuter rails. KIC, however, is cut into two pieces by the major artery Songhu Road, though this does smoothly connect KIC with other parts of Shanghai. The pity is that, currently, the area in between Phase One and Phase Two is very much dominated by cars. Hopefully the completion of Metro Line 10 can reduce people's reliance on cars to travel to KIC and promote a more pedestrian-oriented community, which can strengthen the connections of the two phases.

Social Cohesion - *"Cultural identity occupies pride of place in discussions of social cohesion."* [6] One point that deserves mention is whether the innovation district's becoming a newly central area of the city is a good tool with which to understand the project. It is noteworthy that this also results in questions about whether innovation districts are really innovative at all. For instance, Silicon Valley has been recognized around the world as the innovation city, and Rive Gauche is starting to gain a reputation as the new, knowledge-based urban enclave of Paris. KIC at certain levels is read as a place where talent gets together as a byproduct of the university city. In this case, more exploration is deserved into how KIC, in physical and abstract aspects, merges itself into the city while also distinguishing itself as a pioneering innovative district.

5 Storper, Michael. *The Regional World: Territorial Development in A Global Economy*. New York, NY: Guilford Press, 1997.

6 Franke, Simon, and Evert Verhagen, eds. *Creativity and the City: How the Creative Economy Changes the City*. Rotterdam: Nai Uitgevers, 2005.

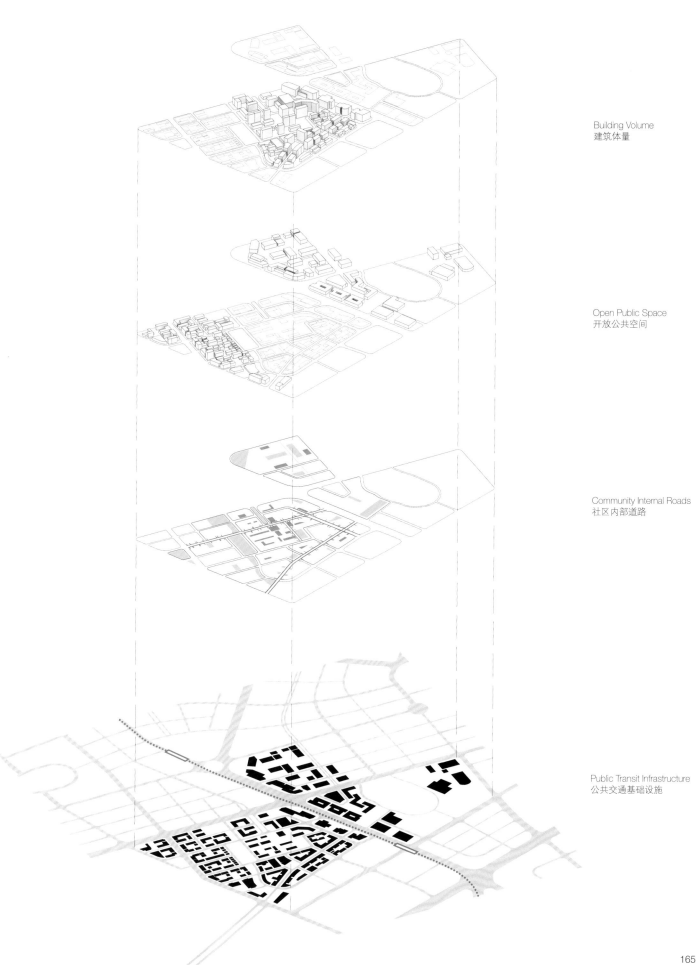

Building Volume
建筑体量

Open Public Space
开放公共空间

Community Internal Roads
社区内部道路

Public Transit Infrastructure
公共交通基础设施

和苹果的校园整体形式，也不同于巴黎左岸的纯公共空间形式，规划师和设计师在创智天地里试图去运用一种截然不同、处于两者之间的形式来创造半围合、半开放、半私密的空间，从而在同一个超大街区中容纳不同的公司。这种介于中间、模糊不清的空间形式却在一定程度上阻碍了其成为一个良好的城市空间。如迈克尔·斯托波所述："*先进的城市经济体的创造者需要有着比模仿的速度更快的创造力才能保持他们领先的位置，才能保证一个更高的生活标准。*"[5] 创新社区应该有更鲜明的定位，要么是作为高度混合的空间和功能的社区，要么就是作为追求高效率和高标准的单一功能的空间。

可持续基础设施 — 可达性也是创新园区的一个主要问题。巴黎左岸、facebook和苹果总部与创智天地位于城市中的不同位置。巴黎左岸位于巴黎市中心，创智天地位于上海市的副中心，而硅谷则距圣何塞有一定距离，离旧金山更远。

新的交通模式似乎使创新园区的设置不再过多地考虑地理位置因素。巴黎左岸的地铁交通和轻轨交通让3万多人在此的工作与生活可以便利地进行。发达的高速路网与许多小型机场和城轨很好地服务于在硅谷工作与生活的人们。淞沪路虽然将创智天地一分为二，但另一方面，担当着连接上海其他区域的主要干道的角色，可惜的是汽车仍是来往创智天地中的主要交通工具。新的地铁完工、地铁站开通后，创智天地的交通模式都发生了改变。很可能人们会选择便利的公共交通从而减少对汽车的使用。创智天地也会成为一个更加友好的步行社区，进而也能更好地加强东西两区的联系。

社会凝聚力 — "文化是社会凝聚力的核心。"[6] 创新园区是否能成为一个城市的新中心是值得讨论的问题，它是否真的能带来创新也是值得注意的问题。硅谷被认为是世界级的创新城区；巴黎左岸从城市飞地到创新园区的地位转变也逐渐得到世界认可；创智天地被认为是人才的聚集地在很大程度上是因为它地处大学城（而非其创新园区的地位）。创智天地必须进一步探索通过抽象和具体的城市设计，来使得它更好地融入城市，同时又保持自己创新先锋的地位。

[5] 迈克尔·斯托波，《区域的世界：经济全球化下的地域发展》，纽约：吉尔福德出版社，1997年

[6] 西蒙·弗兰克与艾福特·费尔哈亨合编，《创新与城市：创意经济如何改变城市》，鹿特丹：NAi出版社，2005年

Fig.10 Aerial View of KIC Phase 1 and 2
图10 创智天地一、二期鸟瞰

Fig.11 Songhu Road as Central Axis of KIC
图11 创智天地中轴淞沪路

KIC is a large community with themes of knowledge and innovation, located in Wujiaochang in Yangpu District. Phase 1 of KIC was planned and designed by SOM with the concept of small blocks and dense streets. In 2011, Tianhua started with planning and designing Phase 2 of KIC, integrating pedestrian-friendly public open spaces like plazas and lawns into the original urban fabric, and distributing public programs along the streets. With this optimized configuration, sports facilities, libraries, and cafeterias promoted interaction between the community and the city, and the academies, hotels, residences, and office buildings along the streets together established a shared community and urban interface. KIC mixed multiple urban functions with open and shared spaces based on universities and innovative industries.

创智天地位于上海杨浦区五角场，是以知识、创新为主题的大型综合社区。创智天地一期由SOM主持规划设计，以小街坊、密路网为主要概念。2011年，天华公司从创智天地二期起开始参与规划设计，在原有路网基础上，将广场、绿地等步行尺度的开放公共空间融入城市环境，并将多种城市功能空间进行整合，将公共功能布置于沿街界面，以体育设施、图书馆、餐厅酒吧等功能增加社区与城市的互动，由街道串联的学院、酒店、住宅与办公组团则营造了共享的社区氛围与活跃的城市界面。创智天地以高校与创意产业为基础，通过开放与共享的理念混合多种城市功能，使自身与周边的城市活力得到提升。

Fig.12 KIC Phase 2 - Open Space and Block Morphology
图12 创智天地二期，开放空间形态与建筑体量

COMPARISONS BETWEEN KIC PLAZA (PHASE ONE) AND UNIVERSITY AVENUE (PHASE TWO) - Mega Block vs Human Scale

KIC combines a number of features to encourage walking and biking. The district is composed of many small blocks connected by narrow streets and plazas, an extension of neighboring city morphology that uses high-quality pedestrian infrastructure. It's a success of the district that the blocks insert seamlessly into nearby city morphology without interrupting the pedestrian experience.

The highly mixed land uses (including many live/work units), strong connection to nearby universities, direct subway access, and a bike-share station all give visitors and residents easy access to walking and biking to run their their daily errands.

Block - The planning strategy of University Avenue has chosen an urban block at a small scale. On the one hand, it responds to the Greater Shanghai Plan, while on the other hand, it is trying to create a more friendly urban experience. Another consideration for choosing the small-scale block—quite different from China's typical city mega block—is the flexibility of creating different courtyard typologies with different functionalities. Within KIC, there are roughly three types of planned blocks: office-dominated areas, such as the KIC Plaza; areas dominated by live-work residential units, such as the University Avenue; and residential areas, such as the gated community in conflict with the original planning idea of the district. The following paragraph will compare the block morphology of KIC Plaza and University Avenue (Fig.10-11) to address their respective design approaches, as well as the impacts and consequences of the actual implementation.

University Avenue is the main pedestrian street at which mixed activities are located. Through an extension of an underground pathway, University Avenue is connected with Fudan University, Jiangwan Stadium, and the center of University City. The size of the block within University Avenue has a width of around 70 meters and a length ranging from 80 meters to 170 meters. At KIC Plaza, in contrast, the size of the block is rather large in scale—ranging from 280 meters by 190 meters (the largest) to 70 meters square. The reason behind the different configurations is of course because of the program distribution of the two areas.

University Avenue is centered on four major concepts (Fig.12-13): the first is to combine live and work spaces all together while to utilize the advantage of nearby universities (such as Fudan University and others), creating an innovative community where the boundary between live and work is blurred.These units are catering to the employees of high tech companies; the second concept is to utilize the water system to create a continuous but formally variable green open space to allow an internal green system, which also serves the greater area outside of KIC. Third is to create a cultural district connecting Fudan University and Jiangwan Stadium. Lastly, by having a human-scale grid where walking and biking is encouraged, the design merges well with the surrounding urban fabric and respects historical architecture.

■ 商业 Commercial	▨ 商业/办公 Commercial/Office
■ 办公 Office	▨ 办公/居住 Office/Residential
■ 酒店 Hotel	▨ 商业/文娱 Commercial/Recreational
■ 托幼 Daycare	▨ 文娱/体育 Recreational/Sports
■ 绿地 Open Space	▨ 商业/绿地 Commercial/Open Space

Fig.13 KIC Phase 2 - Open Space Morphology

图13 创智天地二期，开放空间形态

创智广场（一期）与大学路（二期）比较 – 超级街区对人体尺度

创智天地通过多种手法鼓励步行与骑车。西区由多个小尺度街区构成，其间用较窄的道路连接，延续了场地周边的肌理和高水准的基础设施，所以创智天地西区成功地与周边环境融合并创造了舒适的人行环境。

高度混合的土地用途（如办公、生活、商业等），与附近大学的紧密联系，和已建成的地铁站、自行车租赁站，方便了游客与居民的日常生活。

街区 – 大学路的规划值得一提，它采用了小尺度的城市街区。一方面是对（历史上的）大上海规划的回应，另一方面是为了创造一个友好的人行环境。不同于中国城市中大部分的超级街区，小尺度街区形成的院落能更加灵活地服务于各种不同的需求。总的来说，创智天地存在三种主要的街区类型：以办公为主的创智广场；以工作生活为主的混合住宅商业单元以及与最初的规划理念相违背的以纯住宅为主的封闭小区。文章接下来会对创智天地的大学路和创智广场（图10-11）进行对比，来探讨设计意图以及具体执行的结果。

大学路是西区的主要街道和多种活动聚集的场所，通过一个地下通道，大学路与复旦大学、江湾体育场和大学城中心相连。根据用地性质，大学路周边的街区采用80到180米长、70米宽的小方格网，而创智广场的街区采用了280米到190米和70米宽的较大的方格网。不同的街区网格在很大程度上是为了对应不同的功能需求。大学路片区规划有四个核心概念（图12-13）：一是结合工作与生活并依托大学资源（比如复旦大学和周边的其他高校）创造一个创新社区，将工作与生活融为一体，这里的住宅单元服务于高科技企业的员工和年轻人；其次是以水系为主导创造一系列连续但不同的绿色空间，服务于创智天地和更广大的城市区域；再次是创造具有文化气息的场所，连接江湾体育场和复旦大学；最后是建立以步行和骑车为主的人性化城市空间并与周围环境融合，同时尊重场地历史建筑。

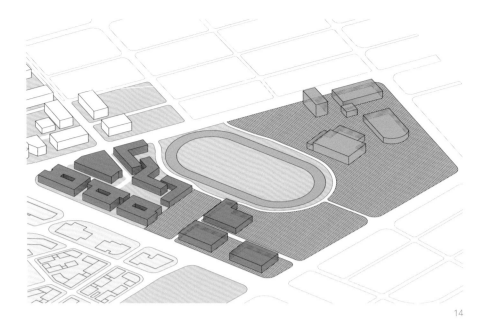

Fig.14 KIC Phase 1 - Open Space and Block Morphology

图14 创智天地一期，开放空间形态与建筑体量

KIC Plaza (Fig.14-15) is centered on four major concepts as well. First is to utilize the existing technology avenue between Keji Avenue and Zhaiyin Road. The design accentuates the function of the technology avenue, while serving as the south gate of the area. Second is to accentuate the connectivity between north Jiangwan City and South University City, while utilizing the planned infrastructure (light rail and subway station). Third is to utilize open space to create a continuous flow across the whole block. Fourth is to create landmark buildings that accentuate the district's identity.

The height of buildings on the block is also influenced by program distribution. The block within University Avenue has a maximum height of 50 meters. Those buildings are distributed around a central plaza and Fudan Plaza. Other blocks are controlled within a height of 24 meters to serve for a comfortable walking experience. At KIC Plaza, building height is controlled within 50 meters closer to Jiangwan Stadium. It grows up to 100 meters further north. As a result, KIC Plaza has a relatively high FAR of 3.1 for offices and 3.5 for hotel/services, while University Avenue has a FAR of 2.0 for live-work and 2.3 for commercial programs.

Because the existing Song Hu Road creates a barrier between the two very distinct areas (University Avenue and KIC Plaza), the two areas have ended up with very different results. The original plan intended to merge the two areas together, mixing the program of KIC Plaza with University Avenue's commercial, residential, and service programs. But the barrier separates the two areas such that one area's program cannot serve the other. As a result, the configuration creates a rather single-use program distribution within each block. KIC Plaza is hard to access from University Avenue. The seemingly mixed blocks in reality are actually two separate areas. Another conflict between the planned block configuration and the actual implementation exists in the northwest area of University Avenue, which today is actually filled with gated communities. Thus the connectivity from Shanghai University of Finance and Economics to University Avenue is interrupted by gated residential programs.

Architecture - The architecture on University Avenue is part of the city fabric; therefore, it should be consistent in terms of style and scale. The frontages of University Avenue are designed to increase the pleasure of the walking experience. Plantations and trees provide cover for pedestrians, while colorful canopies and signage add another layer of richness. The upper levels of buildings are designed with modern aesthetics and details. Utilizing different windows, balconies, and other details helps the buildings serve different functions and programs, as well provide different experiences for pedestrians.

The loft typology allows for living and working together. The flexibility of its nature allows for the entrepreneur's adaptive use of their space. Lofts can be used for small offices, research labs, or galleries. Each of the units occupies two floors in section with an interior staircase for up-and-down circulation. Some of the spaces are double-height, which allows for different uses of the space. If combined with street level parking and commercial spaces, the lofts will provide more flexibility on the lower level. The

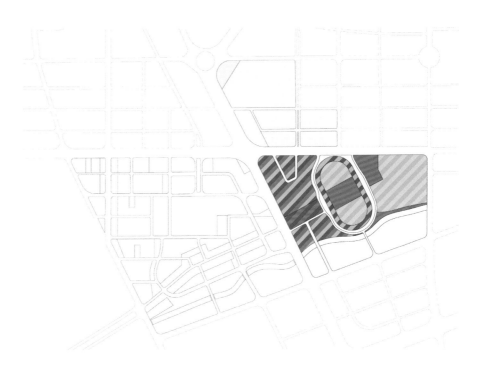

■ 商业 Commercial	▨ 商业/办公 Commercial/Office
■ 办公 Office	▨ 办公/居住 Office/Residential
■ 酒店 Hotel	▨ 商业/文娱 Commercial/Recreational
■ 托幼 Daycare	▨ 文娱/体育 Recreational/Sports
■ 绿地 Open Space	▨ 商业/绿地 Commercial/Open Space

Fig.15 KIC Phase 1 - Open Space

图15 创智天地一期，开放空间形态

创智广场片区规划同样有四个核心概念（图14-15），分别是：利用已有的科技路，强化科技的作用形成区域南入口；利用城铁与地铁，加强江湾体育场与大学城的连接；通过开放空间建立起紧密的内部联系；建立地标建筑加强区域特征。

街区建筑高度根据用途也有所不同：大学路周边的建筑最高为50米，放置于中心广场周边；其他区域的建筑高度控制在24米以内，以便创造舒适的步行环境；创智广场的建筑高度靠近江湾体育场的部分控制在50米以内，向北逐渐增高至100米。所以，创智广场的容积率较高：办公区为3.1；宾馆、服务区为3.5；大学路附近的工作生活区容积率为2.0；商业区为2.3。

由于淞沪路的存在，创智天地被分割为两个不同的区域：原计划中两个区域之间的联系极为紧密，但由于淞沪路的存在和规划中连接两个区域的地下通道并未建成，实际上形成了两个并无太大关联的区域。实施后与原计划的另一个冲突是封闭小区的存在：本应是小尺度的开放街区变成了封闭小区，阻断了原计划中创智天地各地块的紧密联系。

建筑 — 大学路的建筑是城市的一部分，其建筑界面应与城市保持协调一致。临街的建筑旨在提供舒适的人行环境；绿植为行人提供遮蔽，多彩的标志和遮阳棚也为大学路增添了活力；建筑物的上层采用现代美学和细节装饰；不同的阳台与窗一方面服务于不同的建筑功能，另一方面为行人创造了多彩的视觉体验。

小阁楼的居住空间实现了工作与生活的融合。其灵活的空间布局允许小企业家们自发地利用空间，例如小办公室、研究室或画廊等。每一个小阁楼通过内部的楼梯连接两个楼层。一些通高的空间带来了空间利用上更多的可能性。如若与临街的停车场或底层商业结合，小阁楼的空间将会带来更为灵活的空间布局。不同单元服务于不同的人群：家庭、单身公寓或学生等。每个单元都配备有洗衣间、厨房与收发室等。

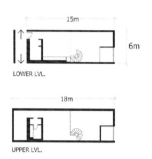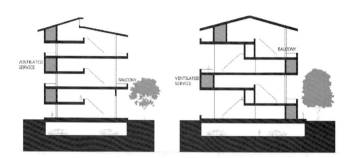

Fig.16 Typical Unit Configuration

图16 典型户型构成

apartments are also designed with different target groups. Unit types include family use, single studio rooms, student dorms, etc. Each of the apartments are equipped with a laundry room, mail services, and kitchens.

The design for large buildings utilize existing open spaces to create a continuous walking experience while also promoting differences in the types of open spaces. Such types includes linear pedestrian belts or enclosed public parks, as well as semi-public parks for office employees. The limitations on building height give the open space a more porous feeling, even in a densely built area.

At KIC Plaza, the architecture serves variable office types. The configuration is designed for flexibility and future expansion. The buildings can be easily rearranged and combined to allow for different scales of office use. The office types refer to Silicon Valley, creating a university-like environment with state-of-the-art infrastructure.

Sustainability was important to attract creative talent to live and work in the district. Shui On Land builds to sustainable standards in most of its developments, and KIC has preliminary registration under LEED (Leadership in Energy and Environmental Design) Neighborhood Development, with individual buildings certified under Commercial Core and Shell at levels from LEED Certified to LEED Platinum, or the Chinese Green Star rating, or both. The certification level of the office buildings has increased over time, with the most recent KIC Corporate Avenue office buildings being pre-certified LEED Platinum. This decision was driven by a desire to make the office spaces stand out in a market projected to be crowded with green buildings and as a statement for the last phase of the project. The buildings feature sophisticated systems to monitor and control daylight, ventilation, and energy use, such as ice-based energy storage, green roofs, and rainwater collection. One project in the development has received RMB 2.5 million in government grants for being a pilot project in energy savings.

But the aim of creating an integrated, mixed-use innovative community is interrupted by the actual implementation. Although the plan is very successful in the design and construction of University Avenue, other streets near University Avenue are not as active; this is partly due to the gated community and partly due to the rather single-use planned area around University Avenue. At KIC Plaza, access is blocked by Song Hu Road, resulting in offices that are disconnected from the residential and commercial programs right next door.

FUNCTIONALITY OF OPEN SPACE - Network vs Central Plaza

Simon Frank and Evert Verhagen in their book "Creativity and the City" claim that the significance of making great spaces with exceptional experiences is that they are crucial to creating a successful innovation district: *"Experience and identity are becoming increasingly important within the physical urban environment. Urban planning and*

Fig.17 University Avenue Street Frontage
Fig.18 KIC Plaza Street Frontage

图17 大学路街道界面
图18 创智广场街道界面

创智天地西区的办公区域利用已有的开放空间形成了一系列连续且不同的线性或围合性的公共、半公共空间以及小型公园等，服务于在此工作的人们。虽然建筑密度较高，但一定程度的建筑高度限制带给人们更加通透的视觉感受。

创智天地东区的办公建筑服务于不断变化的办公要求。各建筑可被灵活地改造或重组，形成不同大小的办公场所。这一灵活布局参考了加州硅谷的校园式街区并采用最为先进的办公设备。

绿色的办公环境是吸引人才生活与工作的重要条件。瑞安地产的大多数项目均采用了绿色的可持续建筑设计，创智天地也通过了美国绿色建筑协会LEED的绿色社区认证。建筑的绿色技术也在不断改善，目前个别建筑甚至达到了LEED白金等级。这一开发决定是基于绿色生态的办公空间可以让社区在与别的办公空间的竞争中独树一帜。这使得创智天地更为突出地体现了先进和可持续的理念，并得到了政府的大力支持，其中的一个建筑甚至获得了政府250万元的拨款支持。

在具体实施中，创智天地也产生了个别不足。比如大学路周边的其他道路由于单一的住宅结构和封闭小区并不活跃，而东区的创智广场则被淞沪路分离形成了一个自发的社区，其较为单一的办公功能与西区的住宅并未很好地融合。

开放空间的功能 – 网状与中心式的开放空间

西蒙·弗兰克与艾福特·费尔哈亨在《创新与城市》一书中提到高品质的空间和卓越的空间体验是一个创新园区成功的必要条件，他们还写道："城市空间的体验与特征正在变得越来越重要。城市规划中的公共空间与空间体验在过去的几十

the design of the public space, and the spatial experience of the city have gained in importance in the last few decades. ... It is not purely functional demands that determine dimension in the design of the environment, for aesthetics and experience-related aspects are just as important." [7]

In this way, at KIC, open space is an irreplaceable layer for making a successful project and organizing the programs and morphologies that can later provide different groups of users with great spatial feelings. The open space here can be categorized into four different types: 1) the public park, 2) the communal space, 3) the street frontage, and 4) the gated open space. For Phase One, located in the eastern area next to the Jiangwan Stadium, the design is laid out according to the large-scale public park as a response to the athletic fields. Phase Two, in the west, is mainly organized by a backbone of streets and a necklace of smaller-scale open spaces, such as communal parks and private gardens. Here the two elements KIC Plaza and University Avenue are the two most typical ones, and also can be read as the most successful elements in the entire project. Along University Avenue all those SOHOs are also regularly organized according to the communal park systems at other dimensions. Due to the layout of this human-scale park, the different programs are well-distributed and the well-designed open space attracts a numbers of users. Adjacent to the SOHO buildings and facing a small central square is a key institution: the Yangpu District government's Overseas Talent building, which helps Chinese who are returning from study and work abroad with residence and work permits and offers its own incentive funding programs for startup businesses. The composition of the park with this key building also promotes the dynamics of the district as a kind of scaled-down Tolbiac block of Rive Gauche.

In Phase One, KIC Plaza and KIC Corporate Avenue together constitute the flagship office and R&D center of the project. KIC Plaza flanks Songhu Road on both sides and is directly adjacent to the Jiangwan Stadium Metro stop. Though Shui-On spent most of their energy on the branding of Phase One rather than Phase Two, it still cannot hide the fact that Phase Two's open space is in the form of network, which works better than Phase One's concentric pattern. We have to admit that the large, single public park next to the stadium successfully fulfilled the task of making a monumental space, but at the same sacrifices the vivid atmosphere created by the small open spaces in a dispersed pattern. Here one important discussion issue that can be raised is the functionality of open space in making a large-scale, monumental open space detached from the public versus creating human-scale and urban-life oriented open spaces. We cannot recklessly argue that the open space setting of Phase Two is better than Phase One; in fairness we should dialectically read them as two correct responses to two different intentions. Then the design problem is pushed to become a social question.

7 Franke, Simon, and Evert Verhagen, eds.
Creativity and the City: How the Creative Economy Changes the City. Rotterdam: Nai Uitgevers, 2005.

年中越来越得到重视……城市规划不再以纯粹的功能为主，美学与感官体验等与功能同等重要。" [7]

从这个意义上说，创智天地中的开放空间是一个无可取代的要素，它们合理地组织功能与空间形态，继而给不同的使用人群一种优越的感官体验。其具体可被细分为四个类型：公共公园、社区空间、沿街界面和半封闭空间。由于创智天地一期东区坐落在江湾体育场旁，所以其公共空间的主要类型是公共公园，作为对其周边环境的一种回应。而二期西区大学路的公共空间是以大学路为中轴，形成了连续的街道和小型公园、社区空间、半公共空间和私有花园等等。大学路与创智广场的开放空间有各自鲜明的特点，可被视为创智天地公共空间营造的典范。大学路附近的"小型办公居住空间"沿社区空间或公园等有效地组织。人性化尺度的开放空间让各种城市功能在其周边得以很好地分布。设计得非常好的开放空间吸引了许多的使用者。在小型办公居住空间旁，面对一个小型中心广场的建筑是杨浦区政府海外人才大厦，为海外学子提供住宅与工作行政上的便利，同时也提供一定的资金支持。这一主要建筑和公园相结合，激活了该区域的城市活力，这与巴黎左岸的托尔比克有相似之处。

创智天地一期的创智广场与企业路一起构成了旗舰办公研究中心。创智天地广场分布在淞沪路两侧，相邻江湾体育场地铁站。虽然瑞安地产将大部分精力花在宣传创智天地一期，但二期西区大学路网状式的开放空间较一期大尺度集中式的开放空间更为活跃。我们也必须承认在江湾体育场前设计大尺度的开放空间，成功地塑造了宏大的标志性空间，但我们也需要意识到（这样的举措）同时放弃了分散的小型开放空间所能带来的活力。这里，有一个关键的关于开放空间功能的议题：是创造单一的大尺度雄伟的空间，还是选择网状式的人性化的小空间？我们不能简单粗暴地认定西区二期大学路片区的公共空间优于东区一期创智广场的空间，在这里必须辩证地看待对创智天地中大尺度开放空间与小尺度开放空间的选择：它们都是对周边的环境以及不同的要求和目的的正确回应。这样就把一个设计问题推向了一个社会问题。

[7] 西蒙·弗兰克与艾福特·费尔哈亨合编，《创新与城市：创意经济如何改变城市》，鹿特丹：NAi 出版社，2005年

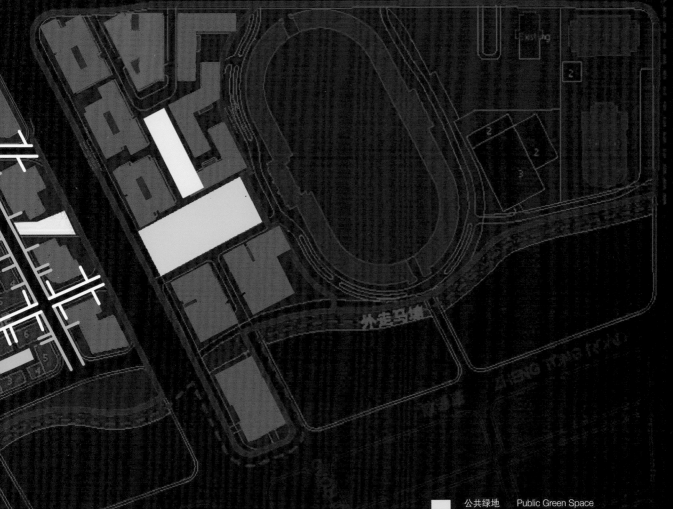

公共绿地 Public Green Space
社区绿地 Community Green Space
建筑体量 Architectural Configuration
街道界面 Street Interface

COMPARISONS BETWEEN KIC AND OTHER URBAN PROJECTS IN SHANGHAI - Development Paradigms

The following paragraph will compare KIC with M50, an arts district in Shanghai, and Xintiandi, a commercial area and successful redevelopment paradigm, in hopes of shedding some light on paradigms of new developments, as well as advantages and shortcomings of the KIC development. These projects all have specific goals that result in different configurations of urban form. The advantages of the KIC project are: 1) KIC is indeed becoming a citywide innovation center and an engine for bringing Yangpu District out of its industrial past. KIC stands as an example of how to turn an industrial center into an engine for innovation; 2) KIC's introduction of the university and the community into the development has enabled it to become a neighborhood in its own right; 3) The old Shanghai Municipal Stadium is an important focal point and landmark for the development, but its reuse presented several design and program challenges; and 4) KIC has also become a magnet for young, creative workers and has put Yangpu on the map as a living and working destination; some local media has even referred to KIC as "Shanghai's uptown."

The KIC Project also has shortcomings: First, the three parts of KIC divided by Zhengli Road and Song Hu Road are not as integrated as was originally planned, bringing the question of whether districts should even be planned as a whole in the first place; second, streets adjacent to University Avenue are not as active as the Avenue itself, making us ask whether the original mixed-use plan was mixed enough or really just superficial; third, the existence of gated communities bring into question the planner's awareness of KIC's real condition, as well as whether there was enough consideration with the actual implementation.

Xintiandi is famous as a redevelopment case in Shanghai (Fig. 19-20). It transformed two blocks of old and dilapidated Shikumen housing into an entertainment center. Compared to KIC, where the development is new, Xintiandi is a paradigm of successful redevelopment. Xintiandi presents a model of adaptive reuse and historic rehabilitation that is making the public accept the value of older housing preservation, especially in an economic transition period. Xintiandi is rightly considered the juncture between preservation and development. By adaptively reusing the dilapidated Shikumen housing, Xintiandi becomes as a window to the past and the future of China, as well as the world. Also, through adaptive reuse, private enterprise could make ideal profits, and the local government could get ideal revenue as well. Xintiandi became a noteworthy leisure and entertainment center, which promoted Shanghai's popularity and increased government revenue. Xintiandi presents a new way for the local government, private enterprise, and international architectural firms to cooperate together, forming a pro-growth coalition, and successfully achieving their goals.

Meanwhile, in its initial stage, Moganshan District was developed for its geographically advantageous position along Suzhou Creek (Fig. 21-22). As a result, nearly all the

Fig.19 Xintiandi
Fig.20 Xintiandi Night View
Fig.21 M50
Fig.22 M50 Night View
Fig.23 KIC Plaza
Fig.24 University Avenue
Source:
https://www.pinterest.com/pin/359936195195004289/
https://www.tablethotels.com/en/shanghai-hotels/the-langham-shanghai-xintiandi
https://stylishheath.com/tag/moganshan-lu/
http://www.panoramio.com/photo/54157955
www.som.com/projects/knowledge_and_innovation_community
http://asiaface.cn/contact.html

图19 新天地
图20 新天地夜景
图21 M50艺术区
图22 M50艺术区夜景
图23 创智广场
图24 大学路
来源：
https://www.pinterest.com/pin/359936195195004289/
https://www.tablethotels.com/en/shanghai-hotels/the-langham-shanghai-xintiandi
https://stylishheath.com/tag/moganshan-lu/
http://www.panoramio.com/photo/54157955
www.som.com/projects/knowledge_and_innovation_community
http://asiaface.cn/contact.html

对比创智天地与上海其他城市改造项目 – 开发项目范例

本文的最后将创智天地与M50（莫干山艺术区）和新天地（商业区与旧建筑改造的典范）相比较，希望阐明成功开发创新区域的条件，同时也讨论创智天地的优势与不足。这几个项目因为开发目的不同，各自均具有独特的形态结构。创智天地的优势在于：第一，创智天地是上海全市的创新中心，同时是带领杨浦区走出旧工业的发动机，是将工业区转化为创新区的典范；第二，引进大学和住宅，使创智天地更像杨浦区的一个邻里，而非外来产物；第三，旧上海体育场是项目中的地标建筑，但在改建中也带来了问题与挑战；第四，创智天地成为创意型人才生活与工作的集中地。一些媒体称创智天地为"上海的上城区"。

同时，创智天地也存在许多不足之处：第一，创智天地的三个主要区域被淞沪路与政立路分割，并未达到规划中理想的完整一体。而本文要问的是，规划中完整的三区联合是否必要？第二，除大学路外，西区的其他街道并不活跃。是否应该进一步加强各种功能的混合，从而激活其他街道？第三，封闭小区的出现是否在规划者的考虑之中？还是规划者对当地地域不了解而造成的后果？

新天地是上海著名的城市改造项目（图19-20）。它成功地将破旧的石库门住宅转变为一个商业中心。与创智天地开发模式不同的是，新天地是在改建中寻找机遇。新天地灵活地利用了历史建筑，并在经济转型时期使大众接受老住宅的价值。新天地也找到了保护与开发的切入点。它成功地改建了历史建筑，成为新旧

factories in the district had direct access to the water. Since this district was used by factories, Suzhou Creek was a waterway rather than a public amenity; the waterfront was inaccessible to the public. Through a series of government interventions and private collaborations, Moganshan District was transformed from a factory and residential area to an arts district. Because of the international organizations and the high-quality art works, it became internationally famous. While some of the artists rent the warehouses for their unique historical and artistic value, others only rent for the low prices. In either case, Moganshan District succeeded in protecting the historic buildings entirely through private efforts, which is a representative bottom-up case for heritage conservation.

In both of the redevelopment cases, there is a strong will from the government, as well as a bottom-up process where public engagement is encouraged. It would have been impossible to successfully develop Xintiandi or the Moganshan District without the intelligence of and collaboration from the public. At KIC, however, public engagement is not part of any of the processes (Fig. 23-24). The resultant gated communities and the conflicts between the original plan and the built reality are therefore inevitable.

对比的窗口。同时，改建也为企业与政府带来利润。新天地随后成为休闲娱乐中心，推动了上海的发展。从开发的角度，新天地是企业、国际建筑设计公司与当地政府合作的典范。

莫干山区因位于苏州河沿岸，对其的开发始于地理位置的考量（图21-22）。因此，所有沿河工厂均可直达苏州河。因其初期作为工业用地，居民并不能直达苏州河沿岸。后期一系列的政府与企业的合作，将莫干山地区转变为一个艺术区并使其获得了国际知名度。一些艺术家被废旧仓库的独特价值吸引而居住于此，另一些艺术家选择这里只因为房租很低。但通过他们的努力，最终保留了这些工业建筑。这是建筑保护中由民间引领、自发保护老旧建筑的典范。

新天地与莫干山艺术区的成功离不开政府与民间的支持。没有公众参与，自下而上的发展就不可能带来这两个项目的成功。而这也是创智天地项目中所缺失的一部分（图23-24）。这也可能是造成封闭小区的一个重要原因。

Context Revitalization
文脉再生

Rainbow City
瑞虹新城

Reconfiguring urban enclaves to mixed-use centralities
重塑城市飞地以创造混合功能新城市中心

Rainbow City

Mixed-Use Urban Redevelopment

Yasamin O. Mayyas

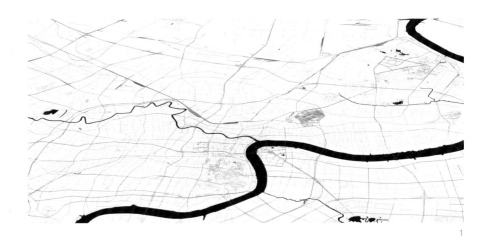

Fig.1 Rainbow City's Location in the Inner City

图1 瑞虹新城在上海内城的位置

Rainbow City was built on the premise of developing an integrated, mixed-use residential community in the inner city of Shanghai. The development aims at creating a planned community that caters to the upper middle class in order to revitalize the existing neighborhood. The community is meant to be supported by its own amenities, creating what can be described as a city within a city, based on the property-led development model dominated by one developer: Shui On Land. This paper aims at analyzing how this model was manifested in Rainbow City by investigating the set of actors involved and the implications for current residents and the existing community. In addition, it tries to understand the relationships to precedents such as Xintiandi, from which certain development strategies were reapplied to Rainbow City, in order to understand what Rainbow City represents in this framework, and to what extent it was successful in achieving its objectives, as well as reevaluating its effects on the socioeconomic milieu.

Mixed-use development has been increasingly adopted by both real estate developers and urban planning departments in Chinese cities. This planning tool has been widely used in European and American cities to reduce the segregation produced by functional zoning. The main objective of the mixed-use development is to create a compact, walkable community through intermixing functions of commercial, residential, and recreational land uses in order to revitalize the socioeconomic status of the community[1], by creating "live, work, play" environments and minimizing commuting distances and reliance on automobiles. Rainbow City is within the city center of Shanghai (Fig.1), anchored by the Bund, North Bund, and Lujiazui business districts, in close proximity to what's known as "the golden triangle," (Fig.2). It aims to create the largest international community hub within Shanghai's inner city ring. The main planning principle is to allocate a commercial street and a green axis, both of 1 kilometer. In addition, it's very well connected to public transport infrastructure by Subway Line 4 and 2 on either end of the commercial corridor.

The planning and the phasing of the project has been synchronized with the urban development of the Hongkou district, especially the district's infrastructural connectivity. The project is envisioned as part of the "three zones, two valleys, and two housing estates" planning initiative that is proposed by Hongkou officials. The initiative targets

[1] Lynch, Kevin, and Joint Center for Urban Studies. *The Image of the City*. Cambridge, Mass.: M.I.T. Press, 1960.

瑞虹新城

雅萨敏·玛亚斯

混合功能的城市再开发

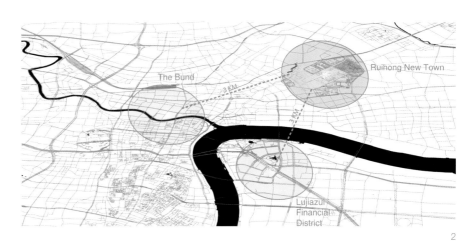

Fig.2 Rainbow City's Proximities

图2 瑞虹新城周边设施

瑞虹新城的建设，是以在上海内城打造综合性混合住宅开发项目为基本概念的，此概念基于由同一开发商控制的地产主导开发模式，为中高收入阶层提供一个精细规划的居住区，并由此创造"城中城"式的、自给自足的社区，重新激活现有邻里。因此，本文尝试通过对瑞虹新城案例的分析，思考此类混合开发模式的实现过程，并通过对此模式中不同参与者的研究，讨论此模式的具体操作方式，及其对现有社区及居民的意义。此外，本文还将讨论瑞虹与其在规划中参考借鉴的先例（如瑞安新天地）之间的关系，从而理解瑞虹在该模式中的定位及其在此框架下实现初衷的程度，同时重新审视瑞虹对该区域社会经济环境的影响。

在中国城市中，混合功能开发策略已被地产商与城市规划部门广泛接受。该策略在欧美城市中作为一种规划手段，被用于减少由城市功能分区（functional zoning）策略导致的城市隔离。混合功能开发策略的主要目的，是通过将商业、住宅与休闲等功能互相融合，创造"居住、生活、休闲"一体化的城市环境，尽可能降低通勤距离，减少对汽车的依赖，从而营造紧凑的、适宜步行的社区，重新激活区域社会经济。[1] 瑞虹新城位于上海市中心区（图1），紧邻由外滩、北外滩与陆家嘴商务区组成的"黄金三角"核心商圈（图2），旨在打造上海内环以内最大的国际化社区。瑞虹新城项目的主要规划要素包括：两条长达1公里的轴线，分别为商业商务轴线与景观绿化轴线；位于商业商务轴线两端的两个地铁站，分属地铁4号线与2号线，保证社区享受便捷的公共交通，同时为社区引入充足的城市人流。

瑞虹新城项目的规划与分期，与虹口区的城市开发进程——尤其是该区域的基础设施连接——达到了高度的协调。该项目是虹口区政府提出的"三区两谷两新城"建设规划的重要部分。"三区两谷两新城"规划旨在优化虹口区功能布局："三区"是北外滩航运与金融集聚区、四川北路商业商务文化休闲区以及以

---

1 凯文·林奇与城市研究联合中心，《城市意象》，坎布里奇，马萨诸塞州：麻省理工学院出版社，1960年

the functional optimization of the district in terms of establishing both a shipping and financial services cluster zone that serves both as Shanghai's international shipping center and financial center, as well as constructing two high-quality housing estates: Rainbow City and the Rainbow Security Housing area. It is also notable that the project is well-served in terms of its proximities, since the district also has two major universities, the Shanghai International Studies University and Shanghai University of Finance and Economics, as well as proximity to some major schools like Shanghai Fuxin Senior High School and East China Normal University Affiliated High School. Planning for the project began in 1996 with the agenda of urban renewal for some of the dilapidated areas in the district. However, in the course of this process it took more than 10 years to relocate the original residents of these areas, some of whom were legal and some illegal tenants, by building offsite housing to compensate and relocate them.

Approval was given to design eight plots without preservation; consequently, this led to demolition operations, which was justified with the claim that it would elevate the poor living conditions and land-use inefficiencies of the old quarters. While the first phase was completed in 2003, only after the second phase was the project accessible by Subway Line 4's Linping Road station. Throughout other phases, residential and mixed-use commercial and leisure amenities were developed simultaneously (Fig. 3). Furthermore, in all the phases the element of mixed-used functionality was incorporated with the aim of commodifying the properties and increasing the property values. Hence one could identify the privatized, exclusive nature of these amenities. As quoted from the website of developer Shui On Land: *"By expanding the scale and variety of retail, office spaces and service apartments through its multi-phased development, Ruihong Tiandi is set to further elevate the commercial influence of Hongkou district."*[2] Incorporating many different commercial-thematic zones along the project's 1 km commercial strip, with the Ruihong Tiandi the commercial spaces shifted from merely targeting the middle class to targetings a high-end audience.

Despite the presence of commercial program in all the project phases, it is still not well-distributed through the blocks. Also, the scale is constant. This doesn't allow for various scales of commercial enterprise to be integrated within the project, which in consequence excludes the opportunity for small business owners to operate in these spaces. Another aspect that I believe should have been considered in the design of the master plan is the block size and its adaptability and flexibility to future and market changes. With mega plot arrangements, it's very difficult to control and direct such development unless it is directed by a central entity, whether that be a developer or a governmental agency.

2 Source: shuionland, http://www.shuionland.com/en-us/property/project/detail/shanghai_ruihongxincheng.

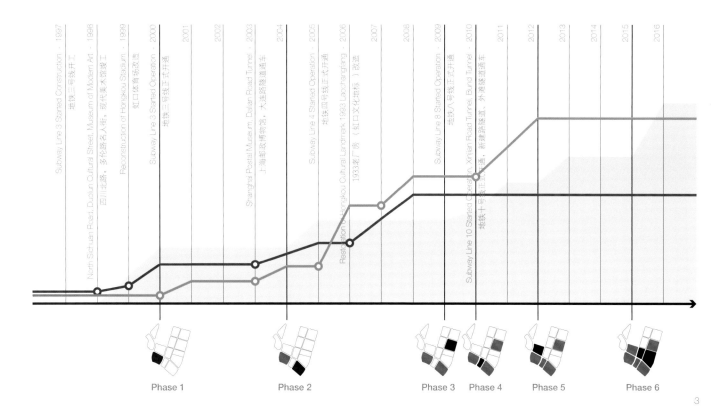

Fig.3 Ruihong Phasing and Planning in Relation to the Urban Development of Hongkou District

图3 瑞虹新城分期规划与虹口区城市开发的关系

大柏树为核心的张江高新区虹口园；"两谷"就是以花园坊为核心的绿色碳谷和上海音乐谷；"两新城"即两个高质量地产项目——瑞虹新城和彩虹湾大型保障性住宅区。另外，值得一提的是，瑞虹新城项目与附近的主要学校联系紧密，例如上海外国语大学、上海财经大学、上海复新高中、华东师范附中等。瑞虹新城的规划开始于1996年，最初的目的是对该区域较衰败的片区进行城市更新。然而，与最初计划不同的是，政府与开发商耗费了超过10年时间，通过在偏远区域建设安置房来重新安置原先社区的合法居民与非法居民。

最终，开发商被允许重新规划设计八个地块，并可以完全拆除而不需要保留原有建筑，从而提高原有场地较差的生活环境条件与较低的土地利用效率。瑞虹新城一期建设于2003年完成，但地铁4号线直到2006年才开通，为瑞虹新城二期引入了大型公共交通站点，即4号线的临平路站。在此后的分期中，地铁站的引入使得居住、商业与休闲设施的同时开发成为可能（图3）。在各个分期中，为最大程度地将地产商品化并提高地产价值，瑞虹最大化地推行了混合功能策略。因此，瑞虹新城项目中的各种配套设施明显带有排外性的私有特征。瑞安房地产的官方网站写道："*通过在多期开发中扩张零售、办公与公寓空间的尺度与多样性，瑞虹天地将进一步提升虹口区的商业影响力。*"[2] 瑞虹新城的1公里商业商务轴线，包含了不同的主题化商业区，其中，瑞虹天地项目将整个商业空间从面向中产阶级转为面向高端用户。

虽然商业功能在瑞虹各期项目中都有分布，但整个街区中商业功能的布局却不够理想，尺度变化也不够灵活。因此，瑞虹新城中的商业规模较为单一，小型企业难以生存。另一方面，地块面积也需要更细致的考虑，以灵活适应未来市场的变化。除非一个开发项目由强有力的组织——不管是开发商还是政府机构——单独主导，否则像瑞虹新城这样的超大地块开发是十分困难的。

2 来源：瑞安集团网站，http://www.shuionland.com/en-us/property/project/detail/shanghai_ruihongxincheng

Fig.4 The Former Residential Neighborhood Fabric before the Demolishment

图4 拆除前的居住区肌理

Moreover, the efficiency of designing for mixed uses at a mega-plot scale is reduced due to the limited street network in relation to the block size and access points. This also limits control in providing different scales of streets, from fully pedestrianized to shared streets. Another important aspect for sustaining a large amount of commercial program is to create a destination that targets customers from across the city to sustain these businesses. It is imperative to understand that the local residents have their own preferences as to where they spend and buy goods; that's why it is important to target a wider audience. This targeting should also be supported by creating a memorable shopping experience that goes beyond simply thematically allocating various commercial activities; instead, a sense of identity and uniqueness should be created, which could be achieved through incorporating and preserving existing activities or structures.

Furthermore, the project was built after Xintiandi and tried to use the same developmental strategies to promote and organize the project. This included, for example, the involvement of international actors in the design team, such as the engagement of Ben Wood as the master plan consultant to brand the project as a cosmopolitan hub. Additionally, the project incorporates the commercial program as the driver and generator of the gentrification process to elevate sales profitability and increase property values. However, the mixed-use program in Rainbow City is not as successful as in Xintiandi. I believe this is due to the deviation from the phasing and development strategy Xintiandi was based on. For Xintiandi, the development strategy was based on buying the land and then developing the mixed-use plots first, creating a landmark in the city. Then, after the first phase was successful and had gentrified the land value and prices around Xintiandi, the rest of the phases were announced and further developed. One could also argue that the success of Xintiandi is due to the element of preservation that referenced a cosmopolitan past and was utilized as part of the marketing strategy. Another major difference is the target group that holds the capacity to sustain the commercial program. For Xintiandi, we can see how the development operates as an attractor at the metropolitan scale, while in Rainbow City the target group is somewhat limited to the upper-middle-class residents who live on the premises of the project itself or within the Hongkou district proper. However, in both cases, the fact that these developments only target certain social groups, from the upper middle class to the upper class, is not sustainable in terms of inclusiveness and social integration. In my opinion, this strategy could fall prey to the pitfalls of social engineering and creating artificial thematic landscapes that lack the vibrancies of incidental and organically evolved areas.

In conclusion, I believe that despite the strides of the mixed-use development, the presence of such large developments concentrated along vast areas of multiple blocks in the inner city presents a significant risk to the existing fabric, both its physical milieu and the socioeconomic milieu as well (Fig. 4-5). It also contributes to creating homogeneous enclaves of people who are from the same socioeconomic stratum, excluding socioeconomic groups with less economic capacity, as well as relocating existing communities in favor of the newly planned one. This can be linked to the development

5

Fig.5 The Former Residential Neighborhood Fabric before the Demolishment, photographed by Geza Radics on Flickr, https://www.flickr.com/photos/radicsge/8398746960/in/album-72157623433509439/

图5 拆除前的居住区肌理
Flickr图片，摄影：Geza Radics，https://www.flickr.com/photos/radicsge/8398746960/in/album-72157623433509439/

同时，超大地块限制了路网密度，降低了规划设计中功能混合的效率，并且减少了从步行路到人车混行路等不同种类道路的可能性。关于保持大量商业业态的另一个重要方面，是为城市其他区域的消费者创造一个消费目的地，从而维持这些商业的生存。对于商业地产开发项目，理解当地居民的消费习惯是必要的，因此开发商需要面向更广的客户群，同时还要创造一种难忘的消费体验，通过保留和纳入现有的商业活动和城市结构带来认同感和独特性，而不仅仅是将不同主题的商业空间分布在场地各处。

此外，瑞虹新城项目以新天地为参考案例，并借鉴其开发模式以提升整个项目的层次，例如邀请国际著名设计师本杰明·伍德，并将其作为瑞虹打造国际化城市中心的品牌推广手段，或通过引入商业功能作为中产阶级化的推动因素，增加商业销售额，提高土地价值。然而，瑞虹新城的混合功能策略并不如新天地那样成功。笔者认为，这是由于瑞虹新城的建设过程与新天地所依赖的分期开发策略存在偏离。对于新天地而言，其开发策略主要是打造城市地标，优先开发混合功能地块，随后在一期开发成功提高周边土地价值的基础上，公布其余的分期开发方案。当然，新天地的成功也可以归因于对老上海国际风格历史元素的保留。另一个主要差异是商业地产项目赖以自我维持的消费群体。对于新天地而言，我们可以看到它如何成为整个城市范围内有吸引力的一个目的地，而在瑞虹项目中，目标群体被限定为居住在瑞虹新城内部或虹口区范围内的高端人群。然而，在两个案例中，商业的目标群体都只面向中高收入群体，这对于社会包容与融合是不可持续的。在笔者看来，这一策略很可能因为其社会工程的圈套与缺乏偶然性与活力的人工主题景观而失败，因为这种偶然性与活力是有机演化的城市空间才能产生的。

作为结论，笔者相信，虽然瑞虹在混合功能开发方面迈出了一大步，但在上海城市中心区进行如此大规模的多街区开发项目仍给现有的城市物理环境与社会经济环境带来了巨大的危机（图4-5）：瑞虹新城创造了一个同质化飞地，固化了高收入社会经济群体，同时将消费能力较低的群体排除在外，为了新规划的社区而

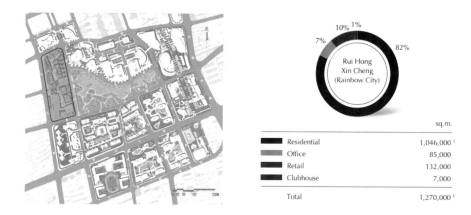

Fig.6 Gross Floor Area Distribution of Xintiandi Project
图6 新天地项目净开发面积组成

model of property-led development, which inherently creates a conflict between the common welfare and the economic targets of the developers. One could also argue that this type of development has been encouraged from the 1990s onwards. Through demolition and relocation, regulation has shifted from compulsory onsite relocation to offsite relocation and compensation. Meanwhile, affordable housing for middle-to-low income families has been absent since the beginning of the second upsurge in property development.

Despite the fact that it's privately developed housing, there should be more diversity in the targeted social group to avoid creating rich or homogeneous enclaves in the city that become frontiers of social exclusion. Hence I think providing the element of affordable housing in order to achieve a more inclusive social mix, apart from a merely functional one, is critical to the sustainability of the project in the long run. One of the planning instruments to achieve that social mix could be inclusionary and incentive zoning policies that aim to provide a certain percentage of affordable housing, perhaps 10-20% either in rental or ownership, along with market rate housing. However, it is imperative that providing affordable housing be consolidated and reinforced at the policy and urban planning levels. Also, the aspect of inclusion should be extended to the type of mixed-use functions in the development and how they could incorporate the existing residents' economic capacity by providing a range of these functions that covers a gradient of economic capacities.

Moreover, the fact that the project distributes mixed-use elements in an axial arrangement concentrated along the 1 km strip for commercial and green axes respectively is not sufficient to achieve the optimum level of mixing. I believe that a more even distribution of the mixed-use program across each urban block that the master plan is composed of would ensure that all the streets are activated (Fig. 6-7). It is essential to keep in mind that this should be achieved with considerations of the scale of these mixed-use programs to provide different feelings and gradients of publicness and tranquility. Another aspect is that the project could have a higher-density street network with a good balance of different street sizes, promoting walkability and cycling along with their programmatic diversity. Also, creating a high density of narrow street sizes with close intersections creates a vibrant, safe, and walkable urban landscape. This would also promote higher density across smaller-scale blocks and have direct effects on the grain and diversity of the building morphology, instead of the current model of homogeneous towers in parks layed out within large-scale blocks.

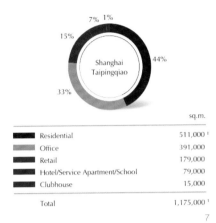

Fig.7 Gross Floor Area Distribution of Rainbow City
图7 瑞虹新城项目净开发面积组成

将原有居民重新安置。这可以与地产主导开发模式联系起来，因为其本质上在公共福利与地产商的经济目标之间制造了冲突。我们也可以说，从1990年开始，拆除安置规定被修改，从强制性原地安置到异地安置与补偿，这种策略就开始被鼓励。同时，中低收入家庭的可支付住宅从第二轮房地产投资风潮开始就一直处于匮乏状态。

虽然瑞虹新城是私人开发的住宅项目，我们仍应该考虑面向更多样化的社会群体，从而避免高收入社区或同质化社区增加社会隔离的程度。因此，笔者认为，为了达到更大的包容性、而非仅仅是功能性的社会融合，提供可支付住宅要素对于维持瑞虹项目的长期可持续性是非常重要的。为了达到这种社会融合，一条规划方面的建议是，通过带有奖励机制的包容性区划政策，提供一定量的可支付住宅单元，通常约为10%-20%左右，向中低收入阶层出租或出售。当然，提供可支付住宅也必须要有足够的规划政策与法律支持。并且，包容性的范围还应该被延伸到混合功能与原有居民消费能力的符合程度，通过提供各种不同层次的商业服务，满足不同消费水平的居民需求。

另外，瑞虹新城的混合功能大多分布于商业商务轴线，相比起来，景观绿化轴线的功能混合程度则相对不足。笔者认为，混合功能在每个地块上更均匀地分布将使得所有的街道都被激活（图6-7）。我们需要时刻记住，这一点需要通过考虑混合功能的尺度来达到，从而提供不同的关于公共性与私密性的感受。另一个方面是该项目本可以获得更高的路网密度，更好的不同道路宽度的平衡，从而提升行人与自行车的通行体验，提高城市功能的多样性，同时创造更多的较窄道路与较近的路口将营造更有活力、更安全的都市行走体验。另一方面，更高的路网密度将带来更小的街区，这将对建筑形态的多样性造成直接影响，避免现有大街区所带来的同质化的"塔楼-公园"模式。

Masterplan
总平面图

	瑞虹新城 Rainbow City	主要水系 Main Waterbody
	周边建筑 Context Buildings	开放空间 Open Space
	主要道路 Main Streets	地铁站点/线路 Subway Station / Lines

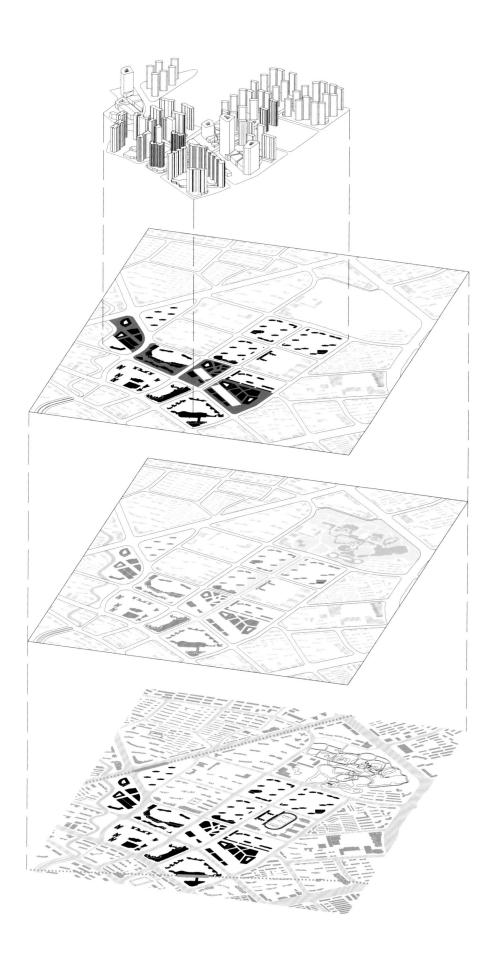

Building Volume
建筑体量:H

Landscape Axis
景观绿化轴线

Commercial / Business Axis
商业商务轴线

Public Transit Infrastructure
公共交通基础设施

Rainbow City is located at Hongzhen Old Street in the old city of the Hongkou District and used to be a crowded, informal settlement. In 2008, Tianhua released a proposal for No. 6 Block in Rainbow City; upon analyzing the context of the site, rather than following the conventional pattern of gated community and street shops, Tianhua opened up the 1 kilometer corridor between the two subway stations and brought in pedestrians to create an urban corridor with commercial, retail, restaurants, and more, promoting the architectural and urban spaces. Shortly afterward, Tianhua cooperated with Benjamin Wood Architects to design the Hongkou Xintiandi complex with restaurants, leisure, and performance spaces. Multiple blocks in Rainbow City have been or are being designed by Tianhua with the purpose of integrating residential programs and urban public space so as to transform the informal housing into a high-end, open, and multi-functional community.

瑞虹新城位于虹口老城区虹镇老街，曾是一处拥挤破败的棚户区。2008年，天华公司首先对瑞虹6号地块进行了设计；经过对周边文脉的分析，不再采用封闭式小区+沿街底商的传统手法，而是将两个地铁站之间长约1公里的路径打开，利用地铁的人流优势创造商业、餐饮、娱乐功能的城市走廊，提升建筑与街道活力。之后，天华又与本杰明·伍德事务所共同完成了虹口天地的设计，结合娱乐、演艺、餐饮等城市功能。天华主持的瑞虹各地块设计都以结合居住功能与城市公共空间为目标，努力创造兼具开放性与复合型的新型社区，完成从棚户区到高品质社区的城市转型。

Rainbow City Development Strategy

Jolene W.H. Lee

Work + Live + Play = Community?

Introduction

Located in the Hongkou district in Shanghai, Rainbow City is one of Hong Kong-based developer Shui On Group's numerous development projects in China. With a total development area of 1.7 million square meters[1], Rainbow City is one of Shui On's key residential parcels within Shanghai's Inner Ring. It enjoys a superior location within the city planning authority's 'Golden Triangle' of key business districts, which is comprised of the Bund, North Bund, and Lujiazui areas.

Located between two subways lines, Youdian Xincun Station on Line 10 and Linping Road Station on Line 4, the developers in their later version of the master plan identified the area as having potential for the insertion of commercial development as an accelerator. When the first phase started in 1998, plans for Subway Line 4 and 10 in the vicinity had not been launched yet – Line 4 started operation in 2005 and Line 10 in 2010. In addition to the subway links, the Dalian Road Tunnel (completed in 2003) was also essential in transforming the connectivity of the Hongkou district to the other side of the Bund. The Hongkou district had also seen massive changes as the years progressed, with key amenities like the Hongkou Football Stadium constructed in 1999 and the activation of key cultural landmarks like the Shanghai Postal Museum in 2003 and the Laochangfang restoration project in 2006.

Sandwiched between three key roads, Dalian Road in the north, Siping Road in the west, and Zhoujiazui Road in the east, the development boundary sits on 13 lots, bounded by Xingang Road, Linping Road, HongZhenLao Road and Quyang Road. Amenities in the vicinity include Peace Park, a municipal park on the northernmost boundary of the development, and numerous schools, post offices, and other civic spaces.

The project itself had been in development for several years, with the first phase launched in 1998 with 10 residential towers and 1,700 units and the latest phase 6 with 4 residential towers launched in 2015 with 680 units (Fig. 1). In total, there had been almost 7,500 units launched since its inception, with the developer gradually adjusting the nuances as they grew more familiar with the demands and gradual changes in the Shanghai property market. The Hongkou district had changed rapidly and so had consumer expectations. The location and transportation networks bounding the development plots were identified to have much greater potential than had originally been planned for.

These rapid changes altered the landscape in the Hongkou district, creating new potentials that the earlier master plan had not anticipated. Initially planned to be completely residential in nature, the developers went through a re-evaluation of the development strategy in 2010 and sought to better leverage these changes through a new master plan, enlisting the assistance of Ben Wood, the architect in charge of what was arguably known as Shui On's flagship project in China, Xintiandi in the Taipingqiao development.

1 Shui On Land Limited. *Strategic Growth: Reaching New Horizons*, (Hong Kong: Shui On Land, 2007).

瑞虹新城开发策略

李文慧

工作+居住+休闲=社区？

概述

瑞虹新城位于上海市虹口区，由香港地产商瑞安集团主导开发，总开发面积约为170万平方米。[1] 瑞虹新城恰好处于由外滩、北外滩与陆家嘴组成的"黄金三角"核心商圈，其地理位置具有巨大潜力，因此，瑞虹新城是瑞安集团在上海内环路范围内最重要的住区开发项目之一。

在瑞虹新城项目的开发过程中，地铁10号线邮电新村站与地铁4号线临平路站先后建成，开发商决定重新规划瑞虹新城，并在新版总平面图中将该区域定义为有能力加速整个区域发展的潜力商业开发区。瑞虹新城一期于1998年开工时，场地周边并无地铁站点，地铁4号线直到2005年才开始运行，地铁10号线2010年才开始运行。除地铁之外，大连路隧道（2003年竣工）也极大地提升了虹口区与外滩对岸的交通联系。近年来，虹口区文化产业同样发展迅速，1999年落成的虹口足球场、2003年的上海邮政博物馆以及2006年的老厂房旧建筑改造项目等，都成为了该地区重要的文化地标。

瑞虹新城被三条重要道路包围：北邻大连路，西邻四平路，东邻周家嘴路。整个项目占据13个地块，周边分别是新港路、临平路、虹镇老路及曲阳路。项目周边设施包括紧邻场地北侧的城市绿地和平公园，以及若干学校、邮局及其他市政服务设施。

整个项目分多期开发，一期于1998年开盘，共由10座塔楼组成，总计1700户；六期于2015年落成，共4座塔楼，总计680户（图1）。整个项目目前总共提供近7500户，在开发过程中，瑞安也在逐渐熟悉上海地产市场的需求与变化，并对开发规模作出相应的调整。虹口区的城市面貌日新月异，本地居民对居住社区的期望也不断提高。事实证明，出众的核心区位与交通便捷度，使瑞虹新城的潜力远远超过瑞安在项目初期的预期。

上述城市建设与发展，从根本上改变了虹口区的面貌，为瑞虹新城带来了初期规划并未预见到的潜在机遇。在开发初期，瑞安对瑞虹新城的定位是仅有居住功能的社区，而在2010年重新评估开发策略之后，瑞安决定更好地利用场地优势，并邀请国际知名建筑师、新天地主创设计师本杰明·伍德（Benjamin Wood）亲自操刀，对瑞虹新城进行了重新规划与设计。

---

1 瑞安集团，《战略增长：展望新视野》，香港瑞安集团，2007年

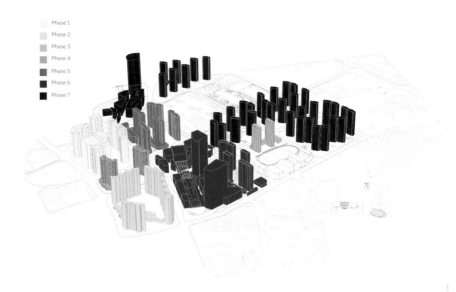

Fig.1 Development Phases of Rainbow City

图1 瑞虹新城项目开发分期

Development Strategy

Rainbow City had been one of Shui On's earliest ventures into the Shanghai market and whilst they had a reputation as a quality property developer from Hong Kong, the earlier Rainbow City phases did not do as well, comparatively, due to their lack of understanding of some uniquely local Shanghai quirks. For instance, the lack of cross-ventilation in units and South-facing main living spaces – cross-air flow and daylight expectations of the local Shanghai populace was something that the developers were not used to, coming from high-density Hong Kong. But beyond that was a lack of understanding of the urban fabric and grain of the larger district and the scale of development, which meant that these early residential development phases resulted in isolated lots of high-density housing with no connection to the street and a lack of continuity and community vitality in the neighborhood.

As consumer expectations changed and they sought a way of living that played on the 'work, live, play' trialectic, the mix of programming in the vicinity of their choice of residence was a key point of consideration (Fig. 2-4). From the developer's point of view, the locality and public transport nodes so conveniently located at the edge of the development boundary was something that could be reconsidered in terms of valuation of their current land assets. The new master plan proposed two key development catalysts, that of a commercial axis and a 'green' axis that cut across the land plots, providing a coherence to the neighborhood.

Taking the two subway stations at either end as markers, the planners designated an axis for commercial development that was planned to allow for seamless integration of pedestrian accessways. This commercial artery was thematised into four main areas: Hall of the Sun, featuring predominantly F&B outlets as main anchors with plenty of international offerings and plenty of alfresco dining areas; Hall of the Moon, identified as the 'centre' of the whole district for culture, recreation, and entertainment programming, with plenty of programmed and unprogrammed spaces to accommodate large-scale events; Hall of the Stars, a family friendly zone with associated programming and retail options; and Fashion Zone, a mega shopping mall with international brands.

The planned commercial artery was based on several studies that identified the positioning of Rainbow City as a commercial centre in the Hongkou district, with due consideration for local and visiting demographics. 85% of consumers were expected to be in the age range of 18-35 years old, mainly young professionals, with almost 40% of these expected to be white-collar executives, and a good majority of them PMETs and college and graduate students from nearby. This demographic was identified to have significant spending power, with a median household income of RMB8000. Nearly 70% were expected to be three-member households, the basic nuclear family unit. Based on these statistics, the commercial artery's main focus group was young adult consumers with high spending power and brand awareness and destination-centric, consumption-

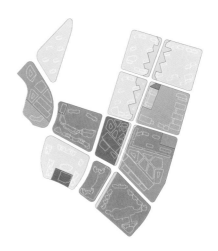

Project Phases
项目分期

一期	Phase 1
二期	Phase 2
三期	Phase 3
四期铭庭	Phase 4
五期璟庭	Phase 5
六期怡庭	Phase 6
未开工	Not Started

Urban Programs
城市建筑功能

商业金融	Commercial
商住混合	Commercial / Residential
生活居住	Residential
文化教育	Educational
社区服务	Community Service
市政设施	Municipal Infrastructure
开放空间	Open Space

Floor Area Ratio
建筑容积率

6.1

2.0

Open Space Ratio
开放空间比率

92%

48%

201

2

3

4

Fig.2 Open Space in Rainbow City
Fig.3 Buildings and Rooftop Open Space in Rainbow City
Fig.4 Buildings and Rooftop Open Space in Rainbow City

图2 瑞虹新城开放空间
图3 瑞虹新城住宅建筑与屋顶开放空间
图4 瑞虹新城住宅建筑与屋顶开放空间

开发策略

瑞虹新城是瑞安集团在上海最早的商业项目之一。然而，虽然瑞安当时在香港已经以高质量的开发项目著称，但瑞安对上海当地市场不甚了解，因此瑞虹新城早期并未取得相应的效果。例如，因为香港超高的城市密度并不能保证穿堂风与南向采光，瑞安在设计瑞虹新城早期住宅时并未考虑南北通风与朝南卧室，而这两者却恰好是上海居民最关心的。而除此之外，瑞安在早期同样缺乏对上海城市肌理与尺度的理解，使得瑞虹新城早期的街区多为孤立的高密度塔楼，与街道缺少联系，邻里活力不足。

随着消费者对住区期望的提高，上海居民开始寻求一种"工作、生活、休闲"三位一体的居住方式，因此，在社区中包含上述的各种城市功能就转化为新住宅开发的重点（图2-4）。另一方面，从开发商的角度看来，公共交通站点可以极大地提升土地价值，因此位于瑞虹新城规划边界的两个地铁站就成为瑞安着重重新考虑的关键。在新规划的瑞虹新城总平面图中，瑞安重点提出了两个激发场地潜力的要素，即商业商务轴线与景观绿化轴线，在重新组织社区空间关系的同时也对场地文脉做出呼应。

瑞虹新城的新规划方案将两个地铁站作为节点，将两个站点之间的一公里线性城市空间作为商业开发空间，为城市提供连贯流动的步行空间。这条商业轴线被划分为四个主题区域："太阳殿"主要为餐饮主题的大型国际美食商业中心；"月亮宫"为区域文化休闲娱乐中心，为音乐表演等大型活动提供空间；"星星堂"为家庭亲子体验商业区；"商业文化连廊"为国际品牌购物区。

在规划该商业商务轴线之前，瑞安针对本地与外来人口进行了大量的市场调研，将瑞虹新城定义为虹口区的商业中心。根据调研，85%的消费者处于18-35岁之间，主要为高学历年轻人才，其中40%为白领人士，多为上海高校的本科或硕士毕业生，从事管理工作。调研认为这一人群具有较强的购买力，家庭月收入中位数为8000元左右，其中近70%为核心三口家庭。基于这一调研结果，商业商务轴线的主要服务对象是具有较强购买力、承认国际品牌、消费观念先进的年轻家庭。针对上海的地理条件与上海市民的行为习惯，该调研还研究了商业区舒适步行距离，并由此确定了两个地铁站之间1公里的商业线性空间规划。

在空间上，步行联系是规划的重点。在瑞虹新城的新规划方案中，规划者着重设计了覆盖地面、地下、空中的步行网络，充分利用了两个地铁站之间的地理优

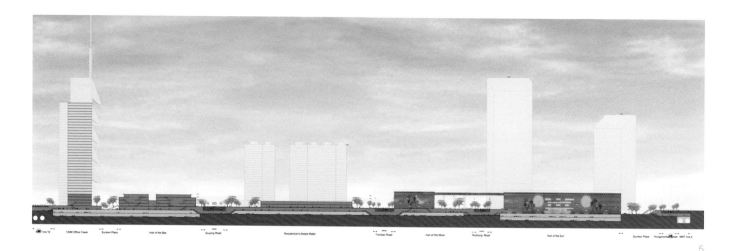

Fig.5 Urban Section of Commerical Axis

图5 商业商务主轴城市断面

savvy young families. The distance between both stations also evens out to an average of 1 kilometer, which the planners had also researched to be an acceptable distance for pedestrianised shopping streets amongst the relevant Shanghai demographic.

Spatially, pedestrian connectivity was key for the planners and they had proposed an extensive network of underground, surface level, and above ground connections, leveraging the connection of the transport network of the two subway stations on either end. In their proposed sections within the commercial spaces' massing, there was to be ample public space and porosity across the building volume to encourage pedestrian circulation from the underground subway links to surface transport options including buses, cycling, and pedestrian-friendly lanes and walkways. (Fig. 5-6)

The green axis runs along Ruihong Road, north-south of the development parcels, linking Linping Road and Xingang Road, leading directly to Peace Park, the main municipal green space in the vicinity. Peace Park is a 260 acre city park which was established in 1958, of which 47 acres are allocated to extensive water features. Ruihong Road was planned such that there would be extensive roadside planting, creating a 'grand avenue' of seasonally appropriate trees and shrubbery, allowing for not just a pedestrian-friendly experience but also encouraging the use and creation of various public spaces and alfresco frontage for the properties.

The planners had also made numerous provisions for curb allowances and setbacks from the road, with cycling lanes and sidewalks with buffer planting in between to reduce noise pollution and increase safety for pedestrians. There were also extensive studies done on traffic speed and carriageway diversions and directions that would contribute to a more pedestrian-friendly experience.

In addition to these surface level provisions for additional green space, the planners had also advised for additional green space to be set aside on commercial properties, encouraging the use of roof gardens and, in the subsequent residential plots, allowed for increased and more varied semi-public green spaces. Sustainability was also a consideration for the updated master plan recommendations and these included numerous active and passive technological strategies and systems which will not be detailed here. Aware of the demographics of their intended market, the planners had sought numerous ways in which the quality of public spaces across the various scales would be heightened without compromising the homeowner's sense of security.

Together, the commercial and green axes were intended to be catalysts for the rest of the development and act as value accelerators for both the developers and the investment-savvy customer base. There have been some marked parallels with Shui On's Xintiandi development, especially with the move of bringing Ben Wood aboard for the master plan update, but there are some key differences with the Xintiandi/Taipingqiao development, which could also be a reason our site observations were markedly different from the expected outcome of the planners.

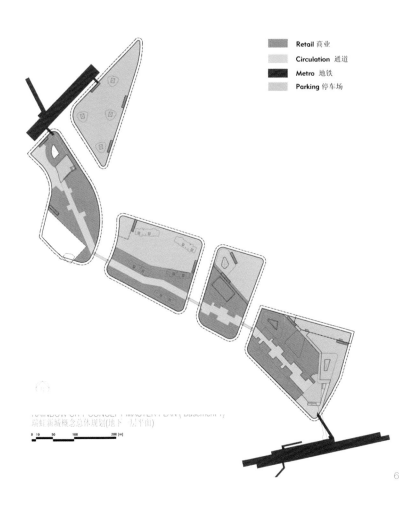

Fig.6 Underground Floor Plan of Commerical Axis
图6 商业商务主轴地下一层平面

势。在方案的商业空间体量剖面中，大量的公共空间穿过建筑，在地下的地铁空间和地面的公交车、自行车及人行道之间建立连贯流通的过渡，为行人创造了舒适的步行体验。（图5-6）

景观绿化轴线沿瑞虹路南北贯穿整个瑞虹新城，将临平路与新港路连接起来，直接通向周边主要的城市绿地和平公园。和平公园建成于1958年，占地264亩（约合17.6公顷），其中47亩（约合3.13公顷）为水景。瑞虹路的规划为瑞虹新城提供了大量行道树，营造了一条绿化随季节变化的"景观大道"，不仅改善了步行体验，也激活了沿路建筑街道界面公共空间的创造与使用。

规划者同时还做出了一系列针对人行道宽度与建筑退线的规定，提供了人行道与自行车道，并利用花坛制造缓冲带，减少噪音污染，保证行人安全。规划者还针对交通流速与道路剖面进行了大量研究，从而更科学地对步行体验进行改善。

除了上述增加地面绿化空间的规定以外，规划者还为商业空间提出了增加绿化面积的建议，提出营造屋顶花园的策略，并鼓励在住宅地块中增加更大、更多样的半公共绿化空间。在对上海市民的行为习惯进行充分研究之后，规划者提出了多种途径来提高公共空间的质量，并避免因此降低业主的安全感。

上述的商业商务轴线与景观绿化轴线将共同激活瑞虹新城的剩余开发项目，对于开发商与希望投资的消费者而言，这两条轴线将成为提升地产价值的主要因素。瑞虹新城与太平桥新天地项目有不少共同点，尤其是两者都邀请了国际著名建筑师本·伍德对整体规划进行调整，但瑞虹新城与太平桥新天地项目仍有一些重要的区别。这也在某种程度上解释了为何笔者的场地调研与规划者的预期结果不符。

Comparative Study with Xintiandi

Shui On's profile within the Chinese property market rose perceptibly with the success of the Xintiandi development within the Taipingqiao district. Located within the Taipingqiao redevelopment area, the scale of the development was comparable with Rainbow City's. However, a key difference would be the level of cooperation between the municipal government and the developer due to some unique historical circumstances of the time, as well as an alignment of redevelopment goals between the two[2]. The actors involved, as well as the level of cooperation, were crucial factors in the different development milestones that characterise each of these developments.

It was 1996 when the '365 scheme', a plan for redeveloping 365 hectares of slums, was launched[3]. The Luwan district government, which presided over one of the busiest and oldest inner city districts in Shanghai, was looking to redevelop an extensive area of old alleys in the Taipingqiao area[4]. Shui On eventually came into an agreement with the district government to cooperate on redeveloping a 52 hectare parcel, which at that time consisted of 23 residential blocks of approximately seventy thousand people. The timeline tagged on to the project was fifteen years, with a projected two to three blocks being redeveloped each year.

Previously part of the French Concession in the 19th-20th centuries, the architecture in this area took on a marked reminder of Shanghai's heritage. Significantly, just to the east of the development parcel was a building where the first Congress of the Chinese Communist Part was held, a site of national historic significance.[5] Taking a step back to understand the context within which Shui On was operating, in 1996 and much of the late 1990s, the property market of Shanghai had been slowing down and Shui On group was also not spared the brunt of the Asian economic crisis. To further exacerbate the slowing down of the development process, there were strict conservation requirements due to the historical nature of numerous structures on site. However, the significance of the CCP Congress Hall in the vicinity of the development also resulted in some wheeling and dealing when the president of Shui On Group proposed the restoration of the CCP building as a gift for the 80th anniversary of the CCP. With all these parameters affecting the nature of the development, the initial foray into the Taipingqiao development was the work on Xintiandi: an elaborate image-raising strategy to increase the reputation and property prices of the Taipingqiao area. Hence between the years of 1999 and 2001, the

2 He, Shenjing & Wu, Fulong, 'Property-led redevelopment in Post-reform China: A case study of Xintiandi Redevelopment Project in Shanghai', *Journal of Urban Affairs* Volume 27: No.1 (2005), p.1-23.

3 Gll, Iker (ed), *Shanghai Transforming: The Changing Physical, Economic, Social and Environmental Conditions of a Global Metropolis*, (Barcelona: Actar, 2008).

4 He, Shenjing & Wu, Fulong, 'Property-led redevelopment in Post-reform China: A case study of Xintiandi Redevelopment Project in Shanghai', *Journal of Urban Affairs* Volume 27: No.1 (2005), p.1-23.

5 Wing, Albert, 'Place Promotion and iconography in Shanghai's Xintiandi', *Habitat International* Vol.30 (2006), p245-260.

对比研究 – 瑞安新天地

上海太平桥新天地项目的成功，使瑞安在中国地产市场声名日隆。新天地项目位于太平桥的城市再开发区域，其规模与瑞虹新城相当。然而，由于当时的特定历史背景，以及城市政府与开发商对于城市再开发的目标完全一致，新天地项目中政府与开发商合作的紧密程度高得多[2]，这也是新天地与瑞虹的主要区别之一。可以说，参与开发过程的角色与各个参与者之间的合作紧密程度，是城市开发项目中的决定性因素。

1996年，上海市政府提出"365危棚简屋"计划（即对共计占地365公顷的危棚简屋进行城市再开发）。[3] 当时的卢湾区作为上海城市中心区最繁忙、最古老的区域之一，尝试对太平桥地区的高密度、宅街道居住区进行改造。[4] 瑞安集团与区政府达成合作协议，对容纳约7万居民、23个居住街区的52公顷土地进行再开发。协议中整个项目的建设时间为15年，预计每年完成2到3个街区的开发。

太平桥地区曾是法国租界，建筑风貌独特，是上海的历史文化保护区。同时，在开发项目的场地东侧，紧邻着中国共产党第一次代表大会召开的会场，是一座国家历史文化保护建筑。[5] 在此，有必要回顾一下瑞安新天地项目的场地文脉。90年代后期，上海地产市场增速逐渐下降，同时，瑞安集团与其他大部分企业一样，都受到了亚洲金融危机的冲击。另一方面，上海市历史文物保护政策十分严格，太平桥场地周边的历史建筑众多，也在一定程度上减缓了新天地的开发过程。然而，瑞安集团提出将出资修复中共第一次大会会场作为中国共产党建立80周年纪念的献礼，这一方案使得紧邻开发场地的中共会场反而成为开发商进行政治博弈的优势。在上述所有因素的共同影响下，新天地项目应运而生，为太平桥地区提供了一幅增加区域知名度、提升土地价值的细致图景。因此，从1999年到2001年，新天地成为太平桥地区最早进行混合功能开发的地块，结合零售、休闲、商业与酒店等业态，与历史文物保护建筑相映

---

2 何深静，吴缚龙，《后改革开放时代中国的地产主导再开发：新天地项目案例分析》，《城市问题期刊》，第27期，2005年第1本，1–23页

3 吉尔·艾克尔主编，《上海城市转型：一个国际都市变化中的物理、经济、社会与环境状态》，巴塞罗那：Actar出版社，2008年

4 何深静，吴缚龙，《后改革开放时代中国的地产主导再开发：新天地项目案例分析》，《城市问题期刊》，第27期，2005年第1本，1–23页

5 阿尔伯特·荣（2007），《上海新天地的场所提升与标志性》，《世界栖居》，2006年，第30期，245–260页

Xintiandi area was the first in the development parcel to undergo extensive redevelopment into a mixed-use district with numerous retail, entertainment, commercial, and hotel facilities, set against a backdrop of restored heritage architecture.[6] The outcome of Xintiandi from the perspective of the developer was essentially attempting to barter and draw into an equation the value of heritage and culture to economic returns through a cultivation of a particular image conducive for property development and its intended audience or clientele.

In addition to the commercial aspect of Xintiandi, the developers also worked with the municipal government to transform a land parcel into Taipingqiao Park, a large-scale green space with a man-made lake. The Luwan district government assisted with the demolition and relocation process for both the Taipingqiao Park and Xintiandi plots, allowing for an accelerated process that helped the developer complete this extensive endeavour in record time. They also assisted with obtaining a favourable policy for the construction of this large, public urban green area and, more importantly, the investment burden was shared between Shui On and the Luwan district and municipal government.[7]

Shui On had come upon a win-win solution: they had the support of the municipal government due to an alignment of development goals and the CCP Congress Hall as a incentive at a larger level in terms of deadlines; they would also be able to wait out the market downturn whilst being poised to cash in on the increased land value that the Xintiandi development would inject into the rest of the area. Property development was employed as a strategy to raise the area's reputation and change its image through physical and functional transformation of the area. Whilst numerous scholars have written criticisms about the perceived heritage value and conservation success of the Xintiandi project, and these are the predominant discussions revolving around Xintiandi in scholarly discourse, that will not be the focus here as this paper attempts to draw parallels between the Taipingqiao development and Rainbow City.

Xintiandi's effect on the entire Taipingqiao development cannot be overstated – it was a resounding commercial success and the property values in the surrounding area rose astronomically. Shui On's strategy of Xintiandi was replicated to varying levels of success in numerous other projects across their China portfolio, creating the dual effects of place promotion and an increase in investment potential. But there are key differentiation factors between the Taipingqiao and Rainbow City developments, which suggest that Shui On's ambitions with replicating the prior successful development strategy in this particular instance may fall short of expectations.

Firstly, the audience of the Xintiandi commercial development and for the Rainbow City commercial development differ sharply. Xintiandi is touted as an iconic destination in Shanghai, leveraging a unique heritage character that the Rainbow City development

6 Wing, Albert, 'Place Promotion and iconography in Shanghai's Xintiandi', *Habitat International* Vol.30 (2006), p245-260.

7 He, Shenjing & Wu, Fulong, 'Property-led redevelopment in Post-reform China: A case study of Xintiandi Redevelopment Project in Shanghai', *Journal of Urban Affairs* Volume 27: No.1 (2005), p.1-23.

成趣。[6] 本质上，从开发商的角度看，新天地项目在历史建筑的文化价值与开发项目的经济回报之间进行了巧妙的交换，通过塑造一种互利共赢的图景来说服甲方，从而引导地产开发。

除了新天地的商业因素，瑞安还与市政府合作，将场地中一个地块改造成一个带有人工湖的大型绿地，即太平桥公园。在太平桥公园与新天地两个地块中，卢湾区政府都对拆除安置提供了协助，加快了开发过程，保证了开发依照原计划按时完成。除此之外，卢湾区政府还帮助瑞安争取到一些关于建设大型公共绿地的有利政策。更重要的是，太平桥公园是由瑞安、卢湾政府以及上海市政府共同出资建设的。[7]

在上述过程中，瑞安提出了一种双赢方案：一方面，通过与市政府的沟通，瑞安明确了两者对于城市再开发的一致目标，并主动出资修复中共会场，从而得到了市政府的支持；另一方面，瑞安在市场的低迷期耐心等待，并对新天地项目对周边地区土地价值的提升一直保持充足的信心。通过物理空间与城市功能的改造，瑞安将地产开发转化为一种提升区域知名度、改善城市面貌的城市发展策略。大量学者都认为新天地的历史文化保护策略颇为成功，而这也是目前学术领域围绕新天地项目的主要观点。但本文试图以新天地为参考案例，在新天地与瑞虹新城之间建立联系，并做出相应的比较。

新天地项目对于整个太平桥地区的开发至关重要，其对于激发商业活动与提升土地价值的作用是难以估量的。之后，瑞安将其在新天地项目中的策略复制到其在中国大陆的其他项目中，并在塑造城市地标与增加投资潜力方面都获得了不同程度的成功。然而，太平桥新天地项目与瑞虹新城项目之间存在着重要的差异，这也表示瑞安在瑞虹新城对新天地策略的复制可能并未达到其预期。

首先，新天地与瑞虹新城商业开发的服务人群存在显著差异。新天地的目标是依靠历史文物建筑，打造上海地标商业区，而这一条件是瑞虹新城所不具备的。同时，新天地的服务对象包括外来人群，如外地游客，或在上海生活的外国居民，瑞虹则并未将这一人群纳入服务对象的范围。虽然瑞虹新城项目还有发展的空间，但无论如何，瑞虹新城项目对于非本地居民都不具备像新天地的历史文物建筑一样的旅游文化价值。[8] 另外，新天地还紧邻淮海路商业区，后者与新天地不尽相同，同时又提供了商业集群效应，消费者愿意将两者作为相邻的目的地，而这也是瑞虹新城所不具备的提升因素。

---

6 阿尔伯特·荣（2007），《上海新天地的场所提升与标志性》，《世界栖居》，2006年，第30期，245-260页

7 何深静，吴缚龙，《后改革开放时代中国的地产主导再开发：新天地项目案例分析》，《城市问题期刊》，第27期，2005年第1本，1-23页

lacks. Whilst Xintiandi has faced numerous criticisms as a heritage conservation project of Shanghai's archetypal Shikumen, these are undeniably a defining characteristic of the area.[8] Xintiandi is a destination for numerous visitors, which include the expatriate and tourist crowd; Rainbow City, on the other hand, was not initially planned for such a demographic. Whilst things could be changed in the latter phases of the development, the fact remains that there is nothing unique about the Rainbow City development for visitors outside the immediate vicinity that allows it to aspire for a 'destination' status. Xintiandi also benefits from latching on to the nearby bustling Huaihai Road's commercial entities, offering a point of differentiation and yet attracting a similar crowd; this is something that Rainbow City lacks as a crowd puller and destination in itself.

Secondly, while Xintiandi's Taipingqiao Park is a joint investment between the municipal government and the developers (since it sits within the development parameters), Peace Park in the Rainbow City project is outside the development parameters and hence a municipal park that the developers have no incentive to fully develop. It only borders the site, and whilst it can be taken as a marketing point in the publicity plans as a welcome green space in densely urbanised Shanghai, the fact remains that it is a dated, aging municipal park, which requires an update and reimagining for the twenty-first century.

Thirdly, and perhaps most importantly, there is a perceived lack of cooperation between the district government and the developer. In the context of Rainbow City, there seems to be a lack of urgency and alignment of development ambition between the owners and the authorities. The stakes are decidedly lower without the pull of the heritage destination in the middle of the old city, but there also seems to be a saturation of similar developments, despite the prime location of Rainbow City. There is an observed lethargy within the development parameters and whilst this may be an unfair judgement seeing how much of the site is still undergoing construction, there is also an observable compromise between the master plan's ambitions and the site conditions.

In a strange contrast to how other cities in the United States cooperate with private developers in the context of transit and land value, where cities harness future land values to pay for infrastructure, these projects in Shanghai are perhaps ones where the developers reap the benefits of transit instead of contributing to said infrastructure. The speed of development in Shanghai ensures this, but it is perhaps also a difference between how infrastructure projects are viewed and accelerated in many Asian economies. Subway and bus transit systems provide plenty of benefits for developments, including but not limited to reduced cost, time, and stress for commuters, cleaner air, and more walkable neighborhoods. The logic between the correlation of public transit and land values is pretty straightforward, but in a Shanghai that develops at a breakneck speed such that the future has to be considered in much-accelerated terms, what does this mean for the future of the community?

8 Ren, Xuefen, 'Forward to the Past: Historical Preservation in Globalising Shanghai', *City & Community* 7:1 March (2008), p.23-44.

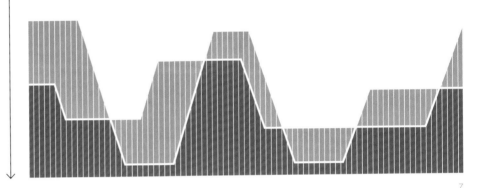

Fig.7 The Various 'Zones' of the Development
图7 开发项目中的多样化"领域"

其次,新天地的太平桥公园处于瑞安的场地范围内,因此公园建设采用了公私合作投资模式,而瑞虹新城以北的和平公园则处于瑞安的场地范围之外,属于公共绿地,因此瑞安并无理由出资建设和平公园。虽然在瑞虹新城的市场营销中,和平公园也作为环境优美的城市绿地成为广告宣传的砝码,但不可否认的是,和平公园建成已久,需要大量资金进行修缮改造,从而适应21世纪的城市生活,而这是瑞安在瑞虹新城项目中并未提供的。

最后,也许也是最重要的一点,是瑞虹新城项目中缺乏地区政府与开发商的协力合作。在瑞虹的开发过程中,两者对于项目目标似乎总是缺少急迫感与一致性。毫无疑问,瑞虹新城没有历史文化建筑的限制,开发的风险被大大降低,但尽管瑞虹新城的区位优越,似乎在周边区域类似的开发项目已经趋于饱和。

这是一组奇怪的对比:在公共交通与土地价值互相作用的背景下,大部分美国城市与私人开发商合作,利用未来的土地价值来为基础设施买单,而这些上海的开发项目则从公共交通中获利,而非为其做出贡献。诚然,上海的城市发展速度使这种策略变得可行,但我们也应从其他亚洲国家的城市公共交通与土地策略中获取参考。地铁与公共汽车系统为城市开发提供了大量机会,包括减少通勤者的资金、时间以及焦虑,减少空气污染,提升邻里步行舒适度等。公共交通与土地价值之间的关系是显而易见的,但在上海这样一个高速发展的城市,"未来"需要以更长远的眼光去看待。那么,对上海而言,"未来社区"意味着什么?

8 任雪飞,向过去前进:全球化上海的城市历史保护,《城市与社区》第7期,2008年3月,23-44页

Work+Live+Play = Community?

The main observation on our site visit was that the development remains disjointed and that there is a lack of the coherence envisioned by the master plan, which is somehow not translating to site outcomes. Whilst efforts had been taken to enliven the neighborhood with the insertion of 'work, live, play' elements to create a district that is self-sustainable and yet sufficiently a destination in itself to attract people outside the development parameters, what it lacks is a palpable sense of community. With all due consideration for the extensive studies and research involved in the proposed master plan, there are some things that the 'hardware' of urban planning and design is unable to provide for. Recommendations could be made by planners on the onset but ultimately, for longevity and sustained development, there needs to be in place a coherent property management manifest that continues long after the last brick is laid. This is the 'software' that any community, to be nurtured into being, depends upon. It involves not just mindset changes on the part of property owners and building management, it also mandates a particular zeitgeist and sense of ownership that extends beyond the four walls of one's apartment and beyond the individual perimeter walls of each residential lot.

In every development, there are numerous actors that catalyze, enable, and facilitate the entire process. But one should never forget the urban neighborhood and always be inclusive in its plans to accommodate the quirks of the urban fabric. What Rainbow City now lacks is the human scale in differentiation across the various spaces, whether commercial, green, or public. Many of the already constructed spaces lapse into a 'generic city' template, where one has difficulty identifying just what it is that makes the neighborhood unique. The master plan considers large swathes of programming, but fails to consider the urban fabric from the perspective of the user. The main commercial and green axes were intended as catalysts, and whilst the actual effect of these remains to be seen, the main criticism of our study group is that these spaces need to be interspersed and integrated much better, with the artificially constructed boundaries between them blurred, if not removed (Fig. 7). It is not just a neighborhood that is being planned, but a community that is to be nurtured; inculcating a sense of ownership and a stake in the larger development is key to this end.

工作 + 居住 + 休闲 = 社区？

在场地调研中，笔者发现整个瑞虹新城的开发仍然处于碎片化状态，各个地块之间联系薄弱，规划总平面图中的整体性并未得到很好的体现。尽管瑞安做出了大量努力，植入"工作、居住、休闲"三位一体的生活方式，提升邻里活力，创造自给自足的社区机制，为非本地居民提供城市目的地，但瑞虹新城仍然缺乏明显的社区感。尽管瑞安为瑞虹新城的整体规划做了大量研究，但城市规划与设计仍然无法提供一些"硬件"。对此，规划者或许可以提出一些初期建议，但更重要的是建立一套健全的管理机制，使得社区在规划建设完成之后仍然可以长期稳定运行。这就是所谓的"软件"，是每个社区赖以持续发展的根本。这种"软件"不仅包含社区业主与管理者思维方式的改变，也需要一种超越私人领域范围的时代精神与主人翁意识。

在每一个开发项目中，有很多参与者激发、帮助并促进了整个过程。但人们不应忘记城市邻里，也不应忘记在规划中涵盖城市肌理的特殊性。现在，瑞虹新城缺少的是在商业、绿化或公共空间中人体尺度的差异性。许多已建成的空间已经陷入一种"通用城市"的范式，人们身处其中时很难辨别该邻里的特殊性。瑞虹的整体规划考虑了整体的功能混合，但并未从使用者的角度思考城市肌理。商业商务轴线与景观绿化轴线被作为社区空间的催化剂，但其实际效果仍有待观察，我们研究小组的主要观点是，这些空间需要更好地穿插与整合，其间的人工边界需要变得更加模糊（图7）。瑞虹不应只是一个经过规划的邻里，更重要的是，它应该成为一个教育和培养市民主人翁意识的新型社区。

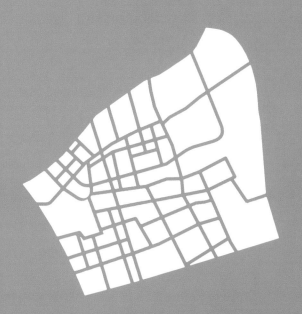

History Reinvention
历史再造

Hengmian Historic Town
横沔古镇

Critical preservation, renovation and rational expansion
批判性保护修复与理性扩建

The Unbearable Lightness of Culture

Reflection on Hengmian Land Development Project

Guan Min

Fig.1. Same Images of Chinese Cities
Source: http://bbs.caup.net/read-htm-tid-34787-page-1.html

图1 中国的"千城一面"现象
来源：http://bbs.caup.net/read-htm-tid-34787-page-1.html

After twenty years of rapid urbanization and development in China, architects and planners are starting to reflect on the homogenous built environment and pay more attention to how to inherit culture and identity of place (Fig. 1). The land development plan in Hengmian, a town designated by the State Council as Chinese traditional settlement in 2013, is a typical project derived from this concern. However, as more and more villages and towns are hastily tagged as historic districts, many scholars have expressed worries about this preservation trend; some have even asserted that preservation has turned into an unwieldy behemoth.[1] Cultural preservation is much more complicated than it seems to be, as many underlying forces invisibly reshape the spatial environment. Therefore, it is necessary to reflect on one question, in terms of place making: what and how should we inherit?

The Homogenous Nostalgia

Hengmian Historic Town is located in a suburban area of Shanghai, approximately equidistant from Pudong New Area and Pudong International Airport. The town was established during the Tang Dynasty, and peaked in the Ming Dynasty as a commercial and cultural center, driven by its developed channel system. With the development of industrialization, many factories moved to Hengmian, yet the industry gradually collapsed as the waterways were replaced by more efficient ground transportation (Fig. 2). Now the historic town is a poor and remote community. Yet what the planned community is trying to delineate is a typical rural image of southern China—dwellings surrounded by rice fields and channels—which isn't the best representation of Hengmian's history and culture. More importantly, this pastoral image is widely applied in many other real estate projects. The key question here is why can this simulated identity be broadly accepted by people with different personal memories and life backgrounds?

As urbanization and modernization have rapidly reshaped the world, people in cities are going through a similar panic about losing their past identities. Nostalgia is this romantic pessimism that prefers things as they are believed to have been and fears the direction of progress today and in the future.[2] So people often choose to mistakenly remember what they want to remember in order to preserve their individual and collective identities. Additionally, people are overwhelmed by public media today. Hence, compared to people's fragile memories, the scenes in movies and the narratives in books provide a clearer and more vigorous image, which finally leads to a collective nostalgia. Meanwhile, the selection of culture is used to establish a distinction from people who have no knowledge or background to appreciate that culture, and hence exclude those people from a given class. For instance, living in this newly proposed Hengmian high-end community not only obtains access to the historic town, but also shows hierarchy and taste. Therefore, ironically, worry about cultural dissipation is facilitating the transition of diverse pasts into simulated memories. And the developers are utilizing this collective nostalgia to organize selective developments and achieve capital accumulation.

1 Goldhagen, Sarah Williams. "Death of Nostalgia." *The New York Times*. June 11, 2011.

2 Lowenthal, David. "Fabricating heritage." *History and memory*, 10.1 (1998): 5-24.

不能承受的文化之轻

闵冠 横沔土地开发项目反思

Fig.2 Old Factory in Hengmian, photographed by Guan Min

图2 横沔老工厂，摄影：闵冠

中国在二十年的飞速发展之后，建筑师与城市规划师开始逐渐反思城市建筑环境的同质化问题（图1），并对于如何传承地域文化和身份愈加关注。作为一个于2013年被国务院评选为中国传统村落的古镇，横沔镇的土地开发项目正是源于这一思考的一个典型案例。然而，随着越来越多的乡镇被政府匆忙地冠以历史区域的名号，越来越多的学者对这一趋势表示了担忧，一些学者甚至将这种保护地区称为"笨拙的怪兽"[1]。由于许多动因在潜移默化地影响着空间环境，文化保护比它表面看起来的要复杂得多。因此在探讨场地营造的时候，除了空间表征，思考文化传承方式和内容是十分必要的。

同质化的怀旧情绪

横沔古镇坐落于上海的郊区，大致位于浦东新区和浦东国际机场中间。古镇建于唐代，在明代时期仰仗其发达的水系成为区域的商业文化中心。随着工业化的发展，许多工厂搬至横沔。但随着水运交通逐渐被更为高效的陆路交通取代，当地工业逐渐凋敝（图2）。如今，横沔成为了一个贫穷而又偏远的地区。尽管该开发项目的规划图景试图描绘一副典型的中国南方农村的景象——民宅被稻田和沟渠所包围，但是这并不能很好地代表横沔当地的历史文化。而更值得注意的是，这种田园景象已经被广泛应用于许多其他地产项目中。这里的核心问题是，为什么各种具有不同的个人记忆和生活背景的人们会广泛认同这种幻象。

随着城市化和现代化迅速改变了世界，城市居民们正在经历相似的、本土身份丢失的恐慌。怀旧情绪在人心中蔓延，这种浪漫的悲观情绪的产生正是由于人们追忆他们相信自己曾经拥有的东西，并且恐惧当今和未来的发展方向。而由于记忆往往很片面和主观[2]，因此人们在记忆他们想要记住的事物以保存他们的个人和集体身份的这一过程中，其得到的结果往往是经过选择的。此外，人们也被当今的公众媒体淹没。因此，相对于人们脆弱的记忆，电影场景和书本叙述提供了更清晰、更有力的形象，这最终导致了共同的怀旧情绪。与此同时，人们也通过文化的甄选来区分他们与那些没有足够的知识背景来欣赏这类文化的人群，彰显自己与其他阶级的区别。举例来说，生活在这个横沔规划高档社区，不仅能够获得毗邻古镇的机会，也能彰显出阶级与品位。因此，讽刺的是，对文化消失的担忧情绪恰恰推动着从多样化的过去转变为集体的虚构记忆。而开发者利用这一集体怀旧情绪去选择性地进行开发，实现资本积累最大化。

1 莎拉·戈德哈根，《怀旧而死》，《纽约时报》2011年6月11日

2 大卫·罗文塔尔，《制造遗产》，《历史和记忆》杂志 10.1（1998）：5-24页

Fig.3. Site Location of Hengmian Old Town

图3 横沔古镇区位

The Enclosed Historic Park and the Extended Capital Flow

Since the world-famous brand Disneyland is constructed right next to Hengmian Historic Town, this whole region has been planned as the new tourism district and has started to attract a huge amount of investment for infrastructure development in the past few years. It was under this circumstance that the Hengmian Land Development project was proposed (Fig. 3-4).

However, the main district of the historic town is excluded from the current development plan, and the future of this enclosed historic area remains undecided. Compared to the history itself, the developers are more concerned that this high-end tourism real estate project first attract rich people from Shanghai and other cities and facilitate the future leisure-tourism industry. Their powerful slogan, "we are constructing a natural, pastoral, nostalgic, and international community," exactly describes the neighborhood that current wealthy citizens dream of.

Yet this project is destroying social patterns and local identity. First, gentrification will push out the local residents, since it is obviously impossible for them to afford the luxury apartments. People who have created and participated in the history will be expelled from their land. Second, the design plan shows that after the high-end community is constructed, there will be no extra space for daily growth and future development; the historic district will be tightly besieged by the new construction. Considering the fact that no alternative industry is planned in the community to support the historic town's livelihood, it is clear that the developer intends to assimilate the historic town into the tourism industry and plan it as a value-added accessory to the high-end community. Finally, regulations in the historic district—caused by its designation as such—will increase the difficulty of pursuing other industries in the fading historic town. It will lead to a preservation that is static in time and partial in picture, all of which undermines the local identity.

Furthermore, the culture of the town will be easily manipulated by tourism developers in the future; the emergence of a homogenous nostalgia has already undermined its fragile history. The developers will utilize the local culture and the cultural brand of "Chinese traditional settlement" to maximize capital profits, which will eventually distort the town's traditional life. Marco D'Eramo expressed his intense worries about the negative effect of UNESCO in his article, "UNESCOcide," where he described the World Heritage listing

3 D'Eramo, Marco. "UNESCOcide." *New Left Review,* 88 (2014): 47-53.

Fig.4. The High-end Community and Enclosed Historic District

图4 高端社区和封闭式历史核心区

封闭的历史公园和扩张的资本流动

由于世界著名品牌迪斯尼乐园建立在横沔古镇旁边，整个区域已被规划为新的旅游区，并已在过去几年吸引了巨额的基础设施投资。在这种情况下，横沔的土地开发项目应运而生（图3-4）。

然而，横沔历史古镇的历史核心区域被排除在目前的发展计划之外，而这个被新开发土地高度围合的历史核心区域的未来仍然悬而未决。相较于历史本身，开发商更关注这种高端旅游地产项目，希望能从上海和其他城市吸引富裕人群，促进未来的休闲旅游产业。他们有力的口号"我们正在建设一个自然、田园、怀旧、国际化的社会"，精确地描述了目前富裕公民的梦想住区。

然而，该项目实际上在破坏社会形态和区域身份特征。首先，这种住宅高档化会将当地居民排除在外，因为这些豪华公寓显然是他们无法负担的。创造和参与历史的人会被驱逐出他们的土地。第二，设计方案显示，历史核心区域将被新建的高端社区紧紧包围，历史区域将没有多余的空间进行日常发展和未来生长。考虑到规划中并没有其他产业来支持古镇的生存，很显然，（在这样的情况下）开发商有意吸收横沔古镇进入旅游业，并计划将其作为一项高端社区的增值附属品。最后，评选为历史区域带来的种种规范将增加在这个衰落的古镇发展其他产业的困难。这些将产生一个在时间上静止、场景上片面的保护规划，最终破坏这里的地域身份。

同质化怀旧情绪的出现已经破坏了脆弱的历史，而城市的文化也很容易被今后的旅游开发者操纵。开发人员将利用当地的文化和"中国传统聚落"这一文化品牌，将资本利润最大化，这将最终扭曲当地的传统生活。德尔茂在他的文章中表达了对教科文组织的负面影响的强烈担忧，他将世界文化遗产名录描述为"*死亡之吻*"[3]。 保护（Preservation）被广泛误读为将场所冻结在一个特定的历史时期，其实这迫使城市切断了每天生长和变化的可能性。同时，联合国教科文组织的招牌诱导着旅游产业通过兜售地域文化来兑现市场价值。巴特尔则用"舞台象

[3] 马尔科· 德拉漠，历史文物的联合国教科文组织化，《新左派评论》，88（2014）：47-53页

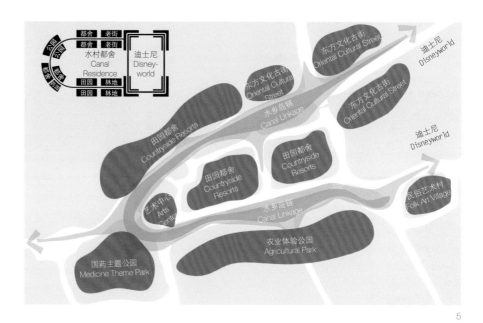

Fig.5. Design Concept

图5 设计理念

as *"the kiss of death"*.³ Preservation was broadly misread as freezing a place in a specific period, forcing the city to cut off the possibilities of daily growth and change. Meanwhile, the UNESCO brand tempted the tourism industry to cash out on the market value by commercializing the identities of places. Diane Barthel summarized these phenomena with the term "Staged Symbolic Communities," which are staged representations of the past or imaginative recreations of generic communities without specific historic referents. *"Any social or economic issues ... are decided not by 1830's residents, but by 1989 officials and curators."* ⁴ The symbolic scene benefits the tourism industry, at the cost of the fabricating local identity (Fig. 5-7).

Pierre Ryckmans argued that although traditional Chinese philosophy shows a different attitude towards preservation, as people pursued immortality by accepting the natural circulation of life and revering enduring thoughts,⁵ China's current preservation policy follows the very western values of UNESCO. This is leading China to become trapped in an even bigger challenge than the western world is facing. However, other than the cultural crisis caused by the enclosure and preservation regulations, it should be realized that in the case of Hengmian the plan to develop a high-end community first and the obvious desire for capital profits will hugely exacerbate the situation. Commerce and tourism are the dominant incentives that utilize and mislead historic preservation. Today, "preservation" is turning into a tool to stimulate tourism and maximize capital flows.

From Morphology to Sociology

In recent years, there has been an apparent trend of Chinese architects paying more and more attention to reinterpreting morphological styles from historic and local construction. Many architects have spread their design practice into rural areas, exploring ways to make their design projects represent local feature and be in harmony with the landscape. It is true that the inheritance of spatial appearance is an important part of culture; however, as mentioned above, many people fail to realize that the growing cultural crisis stems from a sociological unawareness instead of a morphological incomprehension. The bigger problem is negligence toward cultural patterns and social fabric, which eventually dissipates local identity. This is the main reason that Hongcun Village, Charleston, and Hengmian are all suffering from a similar identity loss, even though the preservation of each has seemingly been emphasized.

4 Barthel, Diane. "Nostalgia for America's village past: Staged symbolic communities." *International Journal of Politics, Culture, and Society* 4.1 (1990): 79-93.

5 Ryckmans, Pierre. *The Chinese attitude towards the past*. Vol. 47. Australian National University, 1986.

Fig.6. Hengmian Old Town - Water Village
Fig.7. High-end Residential Community

图6 横沔水乡
图7 高端居住社区

征化社区"来描述这种现象，一种没有具体历史依据的表演化场景表现或者靠想象创造出的通用社区。"*任何社会或经济问题……都不是由1830年代生的居民，而是由1989年的官员和管理者决定的。*"[4] 象征性的场景有利于旅游行业发展，但是付出的代价是真实的地域身份被虚构代替（图5-7）。

莱克曼斯认为，中国传统哲学对于历史保护显示出与西方不同的态度，人们通过接受生命的自然循环、敬畏永恒的思想来追求永恒。[5] 但是中国目前的保护政策是继承自联合国教科文组织的西方价值观，这导致中国陷入一个比西方世界所面临的更大的挑战。而且，除了封闭化历史区域和保护条例带来的文化危机，应该认识到，在横沔案例中，建设高端社区这一首要规划目标和其明显的追求资本利润的欲望会让情况迅速恶化。商业和旅游业是利用和误导历史保护的主要诱因。如今，"保护"的目的，正在变成一种刺激旅游、最大化资本流动的工具。

从形态学到社会学

近年来，中国的建筑师们对历史和地域建筑形态风格的转译给予了越来越多的关注。许多建筑师们将自己的设计实践范围扩大到农村地区，探索结合地域特色进行设计并与景观相协调的方式。的确，空间外观的传承是文化的重要组成部分，然而，正如上面提到的，很多人都没有意识到，越来越多的文化危机是对社会学，而非对于形态学的不了解导致的。文化危机背后更大的问题是对文化模式和社会结构的忽视，而这会最终摧毁本土身份认同。这就是为什么虽然历史保护看上去十分受强调，但宏村、查尔斯顿和未来的横沔会遭受类似的文化破坏和身份丢失。

4 黛安·巴特尔，《怀旧美国旧时的村落：分级符号的社区》，《国际政治，文化，社会杂志》4.1（1990）：79-93页

5 皮埃尔·莱克曼斯，《对待过去的中国式态度》第47卷，澳大利亚国立大学出版社，1986年

Fig.8. Wang Shu's Design in Zhejiang Province

图8 建筑师王澍在浙江省的设计

As David Lowenthal asserted, the key to keeping things alive is not the physical form but the genetic continuity.[6] And this continuity relies heavily on local residents, as they are the witnesses and participants of a place's culture and history, which means they inherit the local social patterns and fabric, not only the historic features, but also future possibilities. Therefore, many people have gradually realized that in many cases, preservation should be replaced by the concept of conservation. Culture is too often understood as a static image, while the truth is that social patterns are changing through time; trying to preserve culture by freezing it in a specific historic period is like being condemned to death, and ignoring the fact that culture is vigorous and transforms with changing times and local activities.

The key to conservation is to pursue a social sustainability, to relate and inherit the local identity and social patterns to spatial and material organization--although as new architectural styles and modern technologies emerge, local people should still be able to find an extension of their social fabric and traditional lifestyle (Fig. 8-9).

Beyond realizing the importance of social sustainability, the dilemma of the identity crisis derives from the fact that economic growth is the ultimate goal in development decision-making. In the book "*The Death and Life of Great American Cities*" Jane Jacobs accused modern urban renewal projects of destroying neighborhood life and social relationships, using the post-war planning of New York City by Robert Moses as a case study.[7] However, capital-based infrastructure constructions were the basis for New York to revive and become one of the most prosperous cities in the world today. Capital is the impetus of social development. Obviously, it is unreasonable to deny the important and positive role that capitalism is playing in the current economic world. The question is, who benefits from the economic income, the locals or the developers?

Tourism and commodification can be an effective tool to achieve a cultural, sustainable development plan in areas that possesses unique culture and history, but lack economic income. However, in most situations, the locals don't benefit enough from the tourism industry and are even excluded from their own land. Therefore, we not only have to realize the importance of social sustainability in cultural conservation, but also need to focus on proposing a commercial plan for the locals instead of the developers—in brief, to make a space of people and places rather than a space of capital flows.

6 Lowenthal, David. "Material preservation and its alternatives." *Perspecta* (1989): 67-77.

7 Jacobs, Jane. *The Death and Life of Great American Cities*. New York: Random House, 1961.

现状肌理采集　　　低层建筑肌理　　　多层建筑肌理　　　高层建筑肌理

Fig.9. Spatial Configuration Design in Hengmian
图9　横沔项目的空间结构设计

正如罗文塔尔断言的那样，保存事物生命力的关键不在于物理形态，而在于本土基因的连续性。[6] 这种连续性高度依赖于当地居民，因为他们是当地历史文化的见证人和参与者。这种连续性应该传承当地的社会形态和肌理，即不仅继承了历史特征，也要拥抱未来的可能性。因此，很多人都逐渐认识到，在许多情况下，保护（Preservation）这一概念应该被保育（Conservation）这一概念取代。人们往往将文化理解为静态图像，而事实是，文化是随着时间变化的，所以试图通过将文化冻结在特定的历史时期（往往是历史中的兴盛时期）来保存文化无异于将文化判处予死刑，因为文化是随着时代和当地活动的变化而变化并保持活力的。

保育的关键是要追求社会的可持续发展，将当地身份和社会模式与空间和材料组织继承和联系起来——虽然新的建筑风格和现代科技会不断出现，当地居民还是能够延续符合他们传统的社会结构和生活方式（图8-9）。

另一方面，"经济增长是发展决策的终极目标"这一不争的事实也导致了身份危机的困境。虽然在雅各布斯《美国大城市的死与生》一书中，雅各布斯指责现代城市更新项目破坏了邻里生活和社会关系，并使用罗伯特·摩西的纽约战后规划作为例证。[7] 但是，基于资本的基础设施建设是纽约复兴成为当今世界最繁华的城市之一的基础。资本是社会发展的动力。显然，我们无法否认资本主义在当前的经济世界发挥着重要积极的作用。现在的问题是这些经济收入究竟使谁大为获益，当地人还是开发商？

旅游业和商品化可以是那些拥有独特文化和历史但是缺乏经济收入地区进行文化可持续发展的有效手段。然而，在大多数情况下，当地人并没有从旅游业中获得足够的收益，有时甚至会被排除出自己的土地。因此，我们不仅要认识到在文化保育中社会可持续发展的重要性，还需要将重点放在为当地人（而非开发商）制定商业计划——简而言之，创造属于人民和地域的空间，而不是资本的空间。

6　大卫·罗文塔尔，《古建保护的材料及其替代品》，《瞭望》杂志（1989年）：67-77页

7　简·雅各布斯，《美国大城市的死与生》。兰登书屋，1961年

Epilogue

The morphological design of the Hengmian Land Development project has inherited local architectural features well, and has developed new spatial types to cater to new lifestyles and social demands. However, the ultimate goal of the project is to stimulate the tourism industry and maximize capital accumulation using the brand and fame of Disneyland and the Chinese Traditional Settlement. The high-end community creates a simulated pastoral image, aiming to attract wealthy people from nearby cities who share a homogenous nostalgia and seek a community for leisure and luxury. In this process, the local history of Hengmian as a commercial and industrial center is being wiped out, and the local residents will be excluded by this gentrification project.

This common cultural-crisis phenomenon should be understood in three aspects. First, culture is being commercialized. Many developers are utilizing the strong effect of public media and the increasing sense of cultural crisis to create simulated nostalgia. Furthermore, the consumption of culture is a way to show taste and hierarchy to achieve exclusion. Second, culture is a life entity; hence, it should be taken seriously. We can choose to open for new development if we believe local history isn't worth inheriting. However, preservation or conservation shouldn't be driven as a brand for economic promotion. Finally, compared to morphological design, more concern should be concentrated on social cohesion to achieve cultural inheritance. Meanwhile, economic pursuits doesn't necessarily produce damage to local identity; the main question is whether the economic benefit is for the locals or the developers.

Culture is more fragile than most people realize, especially in this modern world where everything is commercialized. Hence, exposing and emphasizing the fact of the unbearable lightness of culture is never too much.

结语

横泾土地开发项目的设计形态良好地继承了当地的建筑功能，并发展出了能够满足新的生活方式和社会需求的空间类型。然而，该项目的最终目的是借助迪士尼乐园和中国传统聚落的品牌和名气来刺激旅游业发展并最大化资本积累。高端社区创建了一个虚构的田园形象，旨在吸引在附近的城市里那些怀有相同怀旧情绪、寻求社区休闲和奢侈生活的富裕人群。在这个过程中，横泾曾作为商业和工业中心的历史灰飞烟灭，而当地居民也将被这一中产阶级化的项目排除在外。

这种普遍的文化危机现象应从三个方面来理解。第一，文化正在被商业化。许多开发商正在利用公共媒体的强烈影响和日益增长的文化危机感使他们建构出的怀旧情绪茁壮生长。此外，文化消费也被用作一种显示品位、阶层并实现排外的方式。第二，文化是一个生命体，因此应该被认真对待。如果我们相信当地的历史不值得继承，我们可以选择接受新的发展。然而，无论保护还是保育不应该让村镇被当做一个品牌来寻求经济发展。第三，要想实现文化传承，相对于形态设计，我们更应当关注社会凝聚力。同时，追求经济发展并不必然产生对地域认同的损害，主要问题是能否体察真正的受益者和受害者。

文化比大多数人意识到的更加脆弱，尤其是在这个充斥着商业化的现代世界。因此，无论如何揭露并强调文化的不能承受之轻都毫不为过。

Hengmian is a typical, traditional water town near Shanghai; it was a commercial and trading hub in the era of shipping but declined as modernization prevailed. The newly planned Disneyland just across the street from the town even more substantially crushed the town with pressures and challenges to the historic site and traditional lifestyle. When commissioned to plan for the old town and the extension area, Tianhua came up with the concept of maximizing the historic value and preserving as many historic buildings as possible. They proposed a plan based on waterway crossings and programming different types of residences, public spaces, schools, and waterfront parks along the east-west running river. The plan also extends the island system to the west so as to reshape the community center with a public plaza and buildings and reinforce the connection between the new and the old. Through this proposition, Tianhua is making attempts to not only preserve memories but also meet the contemporary lifestyle.

横沔古镇是上海周边典型江南风格的小镇，在水路交通为主的年代曾作为商品集散地，但随着现代化的进程日渐衰落。而与古镇仅一路之隔的迪士尼乐园的开发给小镇带来了极大冲击，现代文化对传统村镇的压力给区域发展带来巨大挑战。对横沔古镇及其扩展区域的规划，天华带着最大化尊重历史价值的理念，保留大部分历史建筑，围绕十字水街展开规划，沿东西向河道布局不同类型的居住建筑、公共空间、学校以及水岸公园核心区。向西延展出与老街相连的小岛，通过岛上的公共广场与建筑群重塑古镇的新活动中心，强化新区与古镇之间的联系。天华尝试将横沔古镇营造为既保留历史记忆，又符合当代生活要求的新型历史古镇。

Paul Chit Yan Mok
with Dennis Lok Kan Chau

Critical Nostalgia

Outside the long pavilion, along the ancient route,
fragrant grass green joins the sky,
the evening wind caressing willow trees, the sound flute piercing the heart,
sunset over mountains beyond mountains

At the brink of the sky, at the corners of the earth,
my friends wander in loneliness and far from homeland
one more ladle of wine to conclude the little happiness that remains
don't have any sad dreams tonight

- "Farewell," Li Shutong

The world fades away. The noise, the light, the colors, the conversations, and the chaos. Only the bitterness of martinis remains, and a weirdly familiar sense of sorrow that seems transcendent for humanity.

A song emerges from childhood memory. It's hard to tell which emerged first—the melody or the lyrics? The melody is steady and complete while the lyrics have become fragmented and partially forgotten. The theme is clear, nonetheless. It is a theme of immense emptiness and loneliness. It's a song about the sense of getting lost in time. It's a song about nostalgia. "Farewell" was first written by American composer John Pound Ordway in 1851 during the Civil War era. He named the song "Dreaming of Home and Mother." In 1907, Japanese poet Inudo Tamakei wrote the Japanese lyrics and called it "Sorrow of a Traveler." A decade later, Chinese artist Li Shutong rewrote the song again in Chinese and named it "Farewell." Throughout the three versions, the lyrics have taken inspiration from very different local contexts. The lyrics are therefore regional but the melody and spirit are transcendent. The pictures may vary, but the sensation remains the same.

Nostalgia is most significant when it surpasses heavy reliance on pictorial restoration. Such is also the major distinction between "restorative nostalgia" and "reflective nostalgia," as Svetlana Boym points out in "The Future of Nostalgia." When translated into architecture, unfortunately, it is almost too often, that attempts remain at the restorative level. When discussing the commercialization of nostalgia, Boym also brings up a concept from Arjun Appadurai called "ersatz nostalgia," or "armchair nostalgia." It is a form of nostalgia that has no lived experience or collective historical memory. It tricks the consumer into missing what they haven't lost.

Design firm THAPE designed the redevelopment of Hengmian with a clear theme: the field, the water, and the sense of nostalgia. Reviewing the careful spatial reconstruction of a traditional water town, I originally thought this project was closely revisiting the concept of regionalism. But after a closer look at the design brief and presentation, I realized regionalism was at most the means, but never the intention. Instead, it's an intention that pursues nostalgia.

Personally, I think the Hengmian project has pursued to a large extent restorative nostalgia and, when materialized, could be quite successful in delivering armchair nostalgia. To make a more constructive position, or even opposition, in current nostalgic trends in the Chinese architectural scene, however, more thorough thinking could be given to reflective nostalgia.

莫哲昕
及邹乐勤

反思乡愁

长亭外，古道边
芳草碧连天
晚风拂柳笛声残
夕阳山外山

天之涯，地之角
知交半零落
一瓢浊酒尽余欢
今宵别梦寒

——《 送别 》，李叔同

冰凉的夜晚竟让一切融化，世界开始变得朦胧，音乐，影像，对话……融合成一片轻纱似的白雾。唯一清晰的，是杯中渐浓的苦涩，和那仿佛由上辈子遗传下来的一种失落感。

脑际盘旋着一首调子，旋律很完整，歌词却记不清，一哼起来，辞令零碎像杯子表面散聚的水珠。这是一首关于失落的歌，关于穹苍，关于辽阔，关于思念，关于乡愁。《送别》一曲，原为美国音乐家约翰·庞德·奥德威于1851年所作，时值南北战争，原题为《梦见家园和母亲》。1907年日本诗人犬童球溪以此曲为调，重填歌词，更名《旅愁》。又过了十年，李叔同以中文入词，成为《送别》。三首乐曲歌词不一，各自地域背景迥异，所取意象和喻体亦因而有所落差。 然其旋律和意境如出一辙，几乎丝毫不差。

乡愁也是如此，其精髓应当超越了对影像的倚重。在《乡愁的未来》一书中，斯韦兰娜·卜音指出，乡愁可分为"修复型"和"反思型"。修复型乡愁以重新呈现已消逝的影像和物件为主，着重标记性；反思型则探索对怀旧的领会。不幸的是，建筑界对于乡愁的理解似乎从来没有超越过修复型的层面。书中亦提到阿琼· 阿巴杜拉（Arjun Appadurai）有关乡愁商品化的观点，又称"扶手椅式"乡愁。 这是一种缺乏真实经历和集体回忆的怀旧感，是商家透过种种标记，以诱导消费者怀缅他们未曾经历的过去。

天华的横沔古镇重建项目有着鲜明的愿景——"望得见水，看得见田，记得住乡愁"。项目着重研究江南水乡的空间布局并加以仿效，笔者原以为此项目当属于地域主义的范畴，但在细看其设计任务书和项目介绍后，发现地域主义不过是过程，乡愁方为宗旨。

个人认为，横沔古镇作为扶手椅式的乡愁是相当成功的。然而，若要在中国怀旧仿古趋势日盛的环境下探索自身的价值，设计师恐怕要开始对反思型的乡愁多做深思。

Hengmian – A Restorative Nostalgia

Hengmian is located on the outskirts of Shanghai, directly adjacent to the new Disneyland park. In contrast with the nearby international theme park, it is a clear objective of both the developers and designers to celebrate locality and regionalism in this new town development. Hengmian tries to restore an image of an old water town on many levels: from the architectural relationship with the water, to the choice of materials, to spatial and streetscape articulation.

The design team has elaborated on a careful study of sectional relationships between streets, small lanes, and buildings in traditional water towns. The river in the site would serve both a scenic purpose, as well as being a circulation route. The program of buildings adjacent to the water is also calibrated so that when buildings are set back from the waterfront, public events (such as restaurants and cafés) can spill out into the open spaces between buildings and water. There are also times when both sides of the water are enclosed by buildings. The water, in these cases, remains a scene, developing only visual connections with the interiors. Vehicles are allocated along the main roads, which are far away from the water system. The only scenario where land circulation and water circulation overlap is when the scenic walking paths are located next to the water, which is common in traditional water towns. Bridges and small flyovers are also introduced along the water system. The spatial qualities along these bridges (under the bridge, at the two ends of the bridge, etc.) are all designed to formally simulate an old water town.

The major palette of the new town is composed of white powdered walls and grey roof tiles. These are also the most common choices for materials in any old town in China. In the design proposal, woods act as small embellishments between the major architectural masses. Slate stones are the major flooring materials along the walking paths. To mimic an old town, different sizes of stones are collaged into irregular patterns. In a real old town, such collage would be a result of time, due to constant road reconstruction and repairs. It would therefore be a physical manifestation of time and history. These stones would be the elements the village could most closely call "collective memories." In Hengmian's new town, nonetheless, time will not take part in the construction of these patterns. The designers here are restoring a clear pictorial simulation of an "old" town. It is the image that is mimicked, not the process.

In terms of spatial configuration, the design team has defined several distinctive sectional relationships, ranging from the "narrowest" to the "most open." All these sectional articulations are derived directly from the study of an existing historic town. The spatial quality is defined by the heights of buildings and the width of the walking paths between them. (Fig.1) Courtyard houses were scrutinized as the major archetype the design team chose to proliferate across the site. They are categorized into three types: the "enclosed," "semi-enclosed" (three sides enclosed with one side open to street), and

| Alley 极窄 | Street 较窄 | Open Canal Street 开放 |

D/H < 0.5   0.5 < D/H < 1   D/H > 1

Fig.1 Street Dimensional Study in Hengmian

图1 横沔项目的街道空间高宽比

横沔古镇：修复型乡愁

横沔古镇位于上海外沿、新建的迪斯尼乐园旁边。开发商和建筑师的意图甚为明显：要在全球化的主题公园对面，造一个地域的、本土的、属于江南的水乡。这主要是通过（一）处理建筑与水的关系、（二）材质的选择和（三）街巷的空间感来呈现的。

建筑设计团队非常小心地研究了水乡街巷和河道的剖面关系。河流既是主要通道，也是景观所在。每当河水和建筑物不接触时，中间的空地就成为半公共空间，用作露天茶座或餐厅。而当河流两旁均与建筑物触碰时，水则和室内空间构成视觉关系。马路被安置在远离水体的位置。河边的步道是陆路和水路唯一交接的地方。一道道小桥横跨河水，桥底和桥两端的空间都经过反复雕琢，加以深化，以仿效传统水乡的常见布局。

材质方面，新建的横沔水乡和大部分传统江南民居一样，主体采用白粉墙、灰瓦顶，木材作屏风和间墙，地面铺板岩石块。值得注意的是，在传统水乡中，地面的板岩石块会因为历代修复而形成如拼贴一样的不规则图案，是一种历史印记。至于横沔水乡地面的板岩石块砌法，则是设计师为了模仿旧水乡而故意拼贴而成的，是历史标记的一种仿制品。

水乡的空间感主要体现于剖面上的收放自如，其中又以街巷的阔度和两旁楼宇高低的比例至为重要。从"开放"到"窄巷"的空间处理，都是紧随着现有水乡的空间感来安排的。（图1）设计以合院作为建筑的主要类型。合院分三种：封闭式、半开放（三面封闭一面向街）和开放式（两面或以上向街）。和上面提及的地板拼贴一样，传统水乡的不同合院形式，乃是经过时间冲刷而成的。古代一个家族住在一个完整的合院内，合院既是家族内等第的象征，也是社会层级的表现。

沿河生成
EMERGING ALONG THE RIVER

骨架生成
GENERATING SPATIAL STRUCTURE

逐步生长
SPATIAL GENERATION ALONG THE RIVER

老街沿河道交叉口兴起,并逐步沿河呈线性发展。

The Old Streets emerged from the crossing, and has been developed along the river.

河流沿线线性空间继续扩充,同时垂直河道的街巷开始形成,老街基本街巷框架逐步生成。

Linear space expanded along the river, and perpendicular streets emerged, forming the basic structures.

在街巷基本框架下,街区内部建筑逐步填充,线状空间逐步生成面。

The fabrics continued to be filled by buildings based on the basic structure, creating a network of spaces.

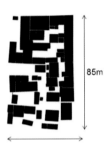
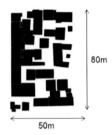
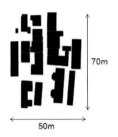
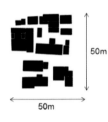

Fig.2 Spatial Generation in Hengmian
Fig.3 Block Typology in Hengmian

图2　横泺项目的空间生成
图3　横泺街区形态类型

"open" (two or more sides open to street). Similar to the slate stones discussed above, such "types" of courtyard houses emerged through time in a traditional town. Most of these traditional courtyard houses were designed to form an enclosed community, serving a very particular family (and hence social) structure that celebrated hierarchy. The semi-enclosed and open "types" only emerged due to reconstruction or demolition of a certain parts of a complete courtyard house. (Fig.2) The designers of Hengmian are mimicking an aging water town. To introduce all three "types" of courtyard houses at the same time is to mimic a finished picture rather than the formation process of an old village. The notion of "old," therefore, remains a restored image rather than an experience of time.

Open squares and "hidden corners" are also introduced along the streets. The squares reassemble the experience of street intersections in older towns while the hidden corners are either dedicated dead-ends of streets or small lanes that serve as detours from main roads. These corners are enriched by planters and the water scenery. The circulation experience of hide-and-seek, or the notion of "found" in a complex, organically grown water town is therefore re-articulated intentionally in this new town design.

The most challenging part for the design team seems to be the articulation of streets and lanes. Not only because they must cater to a range of streetscapes, from roads to small walking paths, but also because the team tries to mimic the naturally-grown street patterns of a historic town. A system always promises regularity. But how shall one design the irregular (Fig. 3)? The designers divided the task into four steps: first, the two main roads are laid out to connect the site with major adjacent highways. Plots are then subdivided into smaller patches. A system of "neighborhood streetscapes" are introduced according to careful study and categorization of found water town streetscapes. Finally, elevations of architecture at the two sides of the lanes are designed to reassemble the circulation experience of an old town. The designers called it the "restoration of traditional streetscape's aesthetics" (Fig. 4).

Aside from a separated zone where modern geometries and façade treatments are applied, through careful study and application of regional architectural materials and colors, as well as a set of calibrated sectional relationships, the old water town spatial experience seems to be quite successfully restored. The scheme is to a certain extent a believable restoration of an old water town, especially on the pictorial level.

Fig.4 Plan of Hengmian Historic Town
图4 横泖古镇总平面

所谓半开放式合院，实为随着历史演变、部分合院重建、拆除，而形成不完整的村落肌理（图2）。而横泖水乡设计团队的三种合院，则是按照今天看见的古老水乡模样复制而成。此乡镇对"古老"的理解是以仿制景象的方式表达的，并不真的牵涉时间和历史。

开放式广场和暗角是街巷的重要元素，前者参考老街交界处，后者则参考老街的尽头和小巷。在天华的横泖项目，水景和种植强化了传统水乡寻寻觅觅的空间感。

设计团队在规划街道上亦颇费心思。一来要把车路、人行道和小径等的空间都规划好，二来还要透过一系列空间法则去模仿古城区不规则的街巷肌理。（图3）团队首先以两条主路贯穿水乡，连接公路，将分割出来的地段细分成小块，然后透过对古城街巷的研究归类，重新编排出不规则的街巷贯穿整个地段。两旁建筑的立面与街巷呼应，以达至设计师"重塑街巷美学"的愿景。（图4）

除了运用现代形式和立面手法处理的区域以外，水乡古镇的总体设计采用了地域性的建筑材料和色彩，塑造了传统的体量关系，由此，横泖项目在对水乡的仿制上可算相当成功。

The Fortune of Nostalgia – Nostalgia as Commodity

But if nostalgia is merely a pictorial resemblance of old building types, how is this new town design different from the adjacent international theme park? Most of the future dwellers of these new houses have never in their lives lived in an old town before. As a matter of fact, based on our on-site investigation, most locals from Hengmian have already moved out to the cities and rented the leftover village houses to the workers of nearby factories. For whom, then, is the nostalgic water town design intended?

Prior to living in this new town, the future residents would never have "missed" living in an older town, for they never lived in one. The nostalgic sensation that the designers attempt to trigger through pictorial restoration therefore lacks a solid foundation: it lacks a lived experience. There are no real memories in the nostalgic feeling.

Such is what Boym would call a "merchandized" or "souvenirized" nostalgia.[1] It is a marketing strategy that tricks the consumer into missing what they haven't lost. As mentioned before, contemporary sociocultural anthropologist Arjun Appadurai first defined such ungrounded nostalgia as "ersatz nostalgia," which means "armchair nostalgia." It is a sense of loss that has no lived experience or collective historical memory.

Svetlana Boym discussed the origin of such commercialized nostalgia in the western context. It was originally derived from Swiss mercenary soldiers who did not want to fight and die away from their homeland. In pop nostalgia (movies or theme parks), nonetheless, the "battle" is a staged one in which the participants can fight on their own terms. Attention to detail is therefore ultimately important, from the choice of guns to the details on uniforms, to make the battle as real as possible. "Everything short of killing." The authenticity here is merely visual, not historical.

The effect is the same in the careful calibration of architectural spaces in the new town design for Hengmian. All the details – from sectional relationships to elevation designs and material choices – are meant to construct a visual symbol linked to a particular part of history that never took place in the consumers' lives.

Boym also brings up a very specific understanding of time in such commercialized, restorative nostalgia. "Restoration signifies a return to the original stasis, to the prelapsarian moment." And such a picture of the past shall not contain any sign of decay. It shall be "freshly painted in its original image and remain eternally young," or, in the case of Hengmian, eternally old. It shall bring us back to the discussion of slatestone pavements and semi-enclosed courtyards. Time is never part of the construction

---

1 Svetlana Boym , *The Future of Nostalgia*, Basic Book, 2001, 46.

乡愁作为商品

但假若横沔的乡愁,仅仅是水乡的景象重塑,那跟毗邻的主题公园又有何分别?这些新房子将来的住客们,很可能这辈子都不曾在水乡生活过,又何来对水乡有所怀缅?实地调研所得,横沔大部分原住民已经搬至市区,现居于横沔的大都是在旁边工厂上班的工人。那么横沔水乡一再雕琢的乡愁,到底是谁的乡愁?

如此情况正好符合卜音和阿巴杜拉提及的乡愁商品化现象。乡愁成为了一种市场策略,去鼓励消费者怀缅不曾失落的过去。所追忆的并不是实有的集体回忆。

卜音在书中探讨了在西方乡愁商品化的成因。[1] 起初,瑞士雇佣兵在战场中产生了厌战的情绪,他们害怕战死沙场,客死异乡。后来,电影和主题公园流行文化为了重现士兵们的乡愁,便将战场情境复制。这些"战场"往往非常迫真,由枪支到制服都一丝不苟,以至让人有"一切极其真实,就只是没有杀戮"之感。这里的"真实"单纯是指视觉上,而非历史上的真实。

卜音对于修复的时间性提出了一个重要的观点。她认为,对于修复型的乡愁,时间是停顿的。这种乡愁必须呈现于"完成"的状态,仿如一幅刚完成的油画,里面画了一切旧世界、旧时代的标记,画一完成,便凝住在那里,恒久不变。横沔重建出来的"古镇"也一样,它并不牵涉时间和过程,而是一件按照"古旧"仿制出来的制成品,在制成的一刻便不应再有所变更。

我们之前提过,水乡的板岩砌石地面和合院空间,在真正的水乡里原是长年累月演化出来的结果。但新建的水乡并没有将时间和过程考虑到筑造当中,而只是将水乡当成是静止的、完成的影像来复制。时间是以标记性的形态呈现的,并不是过程。

卜音于篇末对乡愁商品化作出了如此感叹:"今天,人类文明下的许多产物都能够被大量生产出来。消费者既能享受现代化的便利,亦能继续享受原始的占有欲。娱乐产业(按:包括作为设计师的我们)所提供的怀旧实在是以一种有毒的麻醉剂来慰藉众人心灵的空虚。"

---

[1] 斯韦兰娜·卜音,《乡愁的未来》,基本图书出版社,2001年,46页

process, but it is understood as a frozen, finished, and in a certain way seasoned image. The notion of "old" is never real. It is expressed through symbols and an unnatural collage of materials.

Boym concludes her critiques of merchandized nostalgia with a criticism that I think could be applied to the Chinese architectural trend of nostalgia: "All artifacts of civilization are made available and disposable through mass reproduction; thus the consumer enjoys both the modern convenience and primitive pleasure of fetish possession. Ersatz nostalgia promoted by the entertainment industry [or in our case, by the designers] makes everything time-sensitive and exploits that temporal deficit by giving cure that is also a poison."

Reflective Nostalgia – An Alternative

It is especially hard to define modern Chinese architectural development, due in part to how often we mimic foreign architectural styles. THAPE recently presented five projects the office has undertaken in the past few decades. Some took their style cues from London, some from France, some from Italy. This reveals significantly our "architectural identity crisis." Even when it comes to the Hengmian New Town design, where the theme is one that celebrates locality and regionalism, our means still rely heavily, if not solely, on pictorial restoration (Fig. 5-6).

If nostalgia remains restorative, it will also remain a commodity that does not promise identity. It will forever linger between the terminologies in the lyrics and never reach a melody or spirit that transcends. It questions every metaphor and symbolic vocabulary without establishing a solid theme. When the lyrics are forgotten, the song is too.

As an alternative to restorative nostalgia, Boym suggests the more flexible "reflective nostalgia." While restorative nostalgia strives to restore a picture that is lost (or is claimed as lost), a reflective nostalgia does not focus on the recovery of what is perceived to be an absolute truth but on the mediation of history and the passage of time. While restorative nostalgia evokes a national past and future, reflective nostalgia is more concerned with individual and cultural memory. Time is therefore not expressed as a frozen form, but as a fragmented, composed experience. Rather than a pictorial past that once belonged to the nation, it reflects on the collective culture of the people. And by "culture," Boym refers to the part of human nature that is deeper than collective rituals.

Boym considers the theory of Lev Vygotsky, who believed that beyond the "natural memory" that is closed to perception, humans have a memory of cultural signs that allows meanings to be generated without external stimulation. Remembering, therefore, is not detached from thinking. With the argument of psychologist D. W. Winnicott, Boym suggests that there shall be a "potential space" that would be a manifestation of culture.

Fig.5 Aerial of the existing condition of Hengmian Historic Town

图5 横沔古镇现状鸟瞰

反思型乡愁：另类出路？

中国的现代建筑风格一直难有清晰的定位。某程度上，是由于我们对于模仿外国风格的倚重。我校研究的五个天华项目，有意大利风格的，有英国风格的，有法国风格的。当欧洲风格成为了富贵的象征，彰显的也许是我们建筑界的一场身份危机。我们并没有找到自己的现代化，而是不断模仿外地建筑的形象。来到了横沔重建项目：一个反顾自身的机会，设计师亦理所当然地采用地域主义的手法，然而，其成果却仍然是一件仿制品，只不过，这次我们仿制的是曾经属于自己的风格（图5-6）。

为什么？为什么中国设计师要透过"仿制"来设计中国的建筑？我们"记得住"的乡愁，难道就只是一幅画像而已？难道只是一个标志性的物件？这样的乡愁只是一件便利的商品，并不能回答我们精神上、文明上的身份问题。如果这是一首歌谣，它会徘徊于音乐术语之间，永远达不到超越技术层面的内涵。它难以塑造一个深层次的主题，只是纠结于比喻和象征的手法。当人们遗忘歌词后，歌曲也终将被遗忘。

相对修复型乡愁，卜音提出了另一个概念，即反思型乡愁——其所偏重的不是对旧时代的准确描述或修复，而是对历史的领会，关注的是时间的流逝、个人的记忆和选择。修复型乡愁往往重视一个民族或者宗教整体的过往，时间往往是被渴望停留在历史的某一刻。不同的是，在反思式乡愁的思路里，应该明白时间是流动的，记忆则无可避免地是个人的。对于一个文化体系里面个人选择和集体选择之间的差异，我们应该有更慎重的理解，因为每个人的选择都与其孩提时代的经历息息相关。

Fig.6 Aerial of the existing condition of Hengmian Historic Town

图6 横沔古镇现状鸟瞰

Such space is not defined by any physical forms, but by childhood experience. It is originally the "space" of play between the child and their mother, as that shall be the first space in which the child interacts with the environment and hence develops a sense of "culture". Culture is therefore part of human nature. She therefore concludes that "perhaps what is most missed during historical cataclysms and exile is not the past and the homeland exactly, but rather this potential space of cultural experience that one has shared with one's friends and compatriots that is based neither on nation nor religion but on elective affinities."[2]

Nostalgia therefore does not detach completely from, nor does it rely solely on symbolic resemblance. Reflective nostalgia shall be an intermediary between collective and individual memory. Unlike restorative nostalgia, however, "collective memory can be seen as a playground, not a graveyard of multiple individual recollections." Reflective nostalgia is an ongoing process that does not necessarily demand an actual "homecoming." It contains the elements of both the conscious (symbols) and the subconscious (experience, memories).[3]

To start establishing a reflective nostalgia, aside from formal expression, architecture will have to deal much more intensely with the lifestyles and sensations it provokes. Does it encourage a stronger sense of community? Does it provide space for a different cultural acknowledgement? Or is it still merely an extension of modern urban life, a lifestyle that celebrates individualism and privacy? When Chinese architecture went modern, it was not only the architectural style that was lost. It was the lifestyle, the space, the cultural identity. And the design of Hengmian New Town, just like many new nostalgic architectural proposals, only promises a restoration of a historical image, nothing more. We Chinese architects are still at the brink of the sky, at the corners of the earth, wandering in loneliness and far from our "homeland."

2 ibid, P57.
3 ibid, P46.

她还引用了心理学家温尼科特的理论，认为文化应可以存在于一个比较抽象的、可塑的"空间"里。这个"空间"最开始是形成于婴孩和母亲游戏的空间，孩童成长的过程中，这空间慢慢延伸到他接触到的人和事，从而扩展至社会。在这种理解下，"文化""文明"的定义与人性是密不可分的。她觉得，"在历史的灾难与流放之中，也许，心灵最渴望的并不是过去或真实的家乡，而是这个'空间'，这个跟朋友、同伴共同经历的空间。"[2]

所以乡愁并不单是标记的重塑。反思型的乡愁是建立在集体和个人的回忆之间的："集体回忆应该是一个容许被众人探索的游乐场，而不是少数人怀缅过去的坟墓。"[3] 它应该同时包含了意识（标记）和潜意识（经历、记忆）。

若要探讨反思型的乡愁，建筑设计就不单是形态和风格的问题，要探讨的，是空间对生活的影响，对集体文化的影响，以及对个人记忆、情感的影响。若说要重塑一个"古镇"，我们说的古镇是一种建筑实体吗？是指一种属于过去的建筑类型吗？还是一种意境，一种生活方式，一种文化"空间"？当中国建筑走向现代化，失落的也许不只是风格：还有生活、空间，及文化身份。横沥古镇的重建工程，正如众多怀旧仿古的项目一样，给予了我们标记性的重塑以外，就别无其他。我们设计师依旧站在天之涯、地之角，惘然远眺，还是没有找到回乡的路。

[2] 同上，57页
[3] 同上，46页

PRIMARY REFEREANCES

Archizoom, No-Stop City, 1968-1970. *Domus*, March 1971.

Aureli, Pier Vittorio. *The project of Autonomy: Politics and Architecture within and against Capitalism*. New York: Princeton Architectural Press, 2008.

Barthel, Diane. "Nostalgia for America's village past: Staged symbolic communities." International Journal of Politics, Culture, and Society 4.1 (1990).

Biermann, Christine and Mansfield, Becky. Biodiversity, Purity, and Death: Conservation Biology as Biopolitics, Environment and Planning D: Society and Space 32 (2014).

Bosker, Bianca. *Original Copies: Architectural Mimicry in Contemporary China*. Honolulu: University of Hawai'i Press, 2013.

Boym, Svetlana. *The Future of Nostalgia*. New York: Basic Book, 2001.

D'Eramo, Marco. UNESCOcide. *New Left Review*, 88 (2014).

Den Hartog, Harry, ed. *Shanghai new towns: searching for community and identity in a sprawling metropolis*. Rotterdam: 010 Publishers, 2010.

Denison, Edward., and Ren, Guang Yu. Building Shanghai : The Story of China's Gateway. Chichester, England ; Hoboken, NJ: Wiley-Academy, 2006.

Eco, Umberto, Travels in Hyperreality (New York: Harvest Book, 1986).

Florida, Richard L. *The Rise of the Creative Class : And How It's Transforming Work, Leisure, Community and Everyday Life*. New York: Basic Books, 2004.

Fokdal, Josefine. Bridging Urbanities: Reflections on Urban Design in Shanghai and Berlin. *LIT* Vol. 17. Verlag Münster, 2011.

Frampton, Kenneth. Towards a Critical Regionalism: Six Points for an Architecture of Resistance. Foster, Hal (ed) *The Anti-Aesthetic: Essays on Postmodern Culture*. New York: New Press, 1998.

Franke, Simon, and Evert Verhagen, eds. *Creativity and the City: How the Creative Economy Changes the City*. Rotterdam: Nai Uitgevers, 2005.

Friedmann, John. *China's Urban Transition*. Minneapolis: University of Minnesota Press, 2005.

Gamewell, Mary Louise Ninde. *The Gate Way to China*. New York, Chicago: Fleming H. Revell company, 1916.

Gll, Iker (ed). *Shanghai Transforming: The Changing Physical, Economic, Social and Environmental Conditions of a Global Metropolis*. Barcelona: Actar, 2008.

Goldhagen, Sarah Williams. Death of Nostalgia. *The New York Times*. June 11, 2011

Gregotti, Vittorio. *Inside Architecture*. Cambridge: MIT Press, 1996.

Gregotti, Vittorio. The Form of the Territory. *OASE Journal of Architecture* 80 (2010).

He, Shenjing & Wu, Fulong. Property-led redevelopment in Post-reform China: A case study of Xintiandi Redevelopment Project in Shanghai. *Journal of Urban Affairs*, Volume 27: No.1 (2005).

Hunt, Dixon. *Gardens and the Picturesque: Studies in the History of Landscape Architecture*. Massachusetts: MIT Press, 1992.

Jacobs, Jane. *The Death and Life of Great American Cities*. New York: Random House, 1961.

Jameson, Fredric. Nostalgia for the Present. *Postmodernism, Or, The Cultural Logic of Late Capitalism. Post-contemporary Interventions*. Durham: Duke University Press, 1991.

Katz, Bruce and Wagner, Julie. The Rise of Urban Innovation Districts. *Harvard Business Review*. November 12, 2014.

Leon, Nick. Attract and connect: The 22@Barcelona innovation district and the internationalisation of Barcelona business. *Innovation*. Volume 10 2008 (2-3).

Liu, Yaqing. Shanghai: One City Nine Towns. *Oriental Outlook*. 2015-07-09.

Lowenthal, David. Fabricating heritage. *History and memory*, 10.1 (1998).

Lowenthal, David. Material preservation and its alternatives. *Perspecta* (1989).

Lynch, Kevin, and Joint Center for Urban Studies. *The Image of the City*. Cambridge, Mass.: M.I.T. Press, 1960.

Morpurgo, Guido, and Gregotti Associati. *Gregotti & Associates : The Architecture of Urban Landscape*. New York : Rizzoli, 2014.

Mostafavi, Mohsen. *Ecological Urbanism*. Cambridge: Harvard University Graduate School of Design, 2010.

Ren, Xuefen. Forward to the Past: Historical Preservation in Globalising Shanghai. *City & Community* 7:1 March (2008).

Rutten, Paul. Creativity, innovation, and urban development. *Creativity and the City. How the Creative Economy is Changing the City* (2005).

Ryckmans, Pierre. *The Chinese attitude towards the past*. Vol. 47. Australian National University, 1986.

Sennett, Richard. Borders and Boundaries. Burdett, Ricky and Sudjic, Deyan (eds.) *Living in the Endless City*. London: Phaidon, 2011.

Soria y Puig, Arturo (ed). *Cerdà: the five bases of the general theory of urbanization*. Milano: Electa, 1999.

Storper, Michael. *The Regional World: Territorial Development in A Global Economy*. New York, NY: Guilford Press, 1997.

Tafuri, Manfredo, Vittorio Gregotti: *Buildings and Projects*. New York: Rizzoli, 1982.

Tuan, Yi-Fu. Place: An Experiential Perspective. *Geographical Review* 65, no. 2 (1975).

Urban Land Institute. *ULI Case Studies: Knowledge And Innovation Community*. Washington, D.C. : Urban Land Institute, 2015.

Waldheim, Charles. *The Landscape Urbanism Reader*. New York: Princeton Architectural Press, 2006.

Wilhelm, Miller. *The Prairie Spirit in Landscape Gardening*. Amherst: University of Massachusetts Press, 2002.

Wing, Albert. Place Promotion and iconography in Shanghai's Xintiandi. *Habitat International* Vol.30 (2006).

Xue, Charlie Q.L. and Zhou, Minghao. Importation and Adaptation: building 'one city and nine towns' in Shanghai: a case study of Vittorio Gregotti's plan of Pujiang Town. *Urban Design International* 12, (2007).

Zhou, Min. *Chinatown: The socioeconomic potential of an urban enclave*. Temple University Press, 2010.

主要参考

建筑视窗，《不停留城市》，创作于1968–1970年，摘自《Domus》1971年三月版

皮埃尔·维托利奥·奥雷利，《自治项目：资本主义内与反资本主义的政治与建筑》，纽约：普林斯顿建筑出版社，2008年

黛安·巴特尔，《怀旧美国旧时的村落：分级符号的社区》，《国际政治，文化，社会杂志》4.1期，1990年

克里斯丁·毕尔曼，贝基·曼斯菲尔德，《生物多样性，纯粹性，死亡：作为生物政治的保护科学》，摘自《环境与规划D：社会与空间》，2014年，32期

比安卡·博斯克，《原创复制品：当代中国建筑模仿》，火奴鲁鲁，夏威夷大学出版社，2013年

斯韦兰娜·卜音，《乡愁的未来》，纽约：Basic Book出版社，2001年

马尔科·德拉漠，历史文物的联合国教科文组织化，《新左派评论》，88（2014）

哈利·邓·哈托格（编），《上海新城：追寻蔓延都市里的社区和身份》，鹿特丹：010出版社，2010年.

爱德华·丹尼森，广裕仁，《建设上海：中国门户的故事》，奇切斯特，英格兰；霍博肯，新泽西：威利学院，2006年

乌姆贝托·艾科，《超现实旅行》，纽约，哈维斯特书店，1986年

理查德·佛罗里达，《创意阶层的崛起及其改变工作、休闲、社区与日常生活的方式》，纽约，基本书店，2004年

约瑟芬·弗克达尔。《连接城市：对上海和柏林城市设计的反思》，《LIT》第17期，明斯特出版社，2011年

肯尼斯·弗兰普顿，《迈向批判性地域主义：建筑抗御的六点提议》，摘自《反美学：后现代主义论文集》，哈尔·福斯特编著，纽约，纽约出版社，1998年

西蒙·弗兰克与艾福特·费尔哈亨合编，《创新与城市：创意经济如何改变城市》，鹿特丹：NAi出版社，2005年

约翰·弗里德曼，《中国城市转型》，明尼阿波利斯，明尼阿波利斯大学出版社，2005

玛丽·路易斯·宁德·加姆威尔，《中国门户》，纽约，芝加哥：弗莱明 H. 里维尔公司，1916年

吉尔·艾克尔主编，《上海城市转型：一个国际都市变化中的物理、经济、社会与环境状态》，巴塞罗那：Actar出版社，2008年

莎拉·戈德哈根，《怀旧而死》，《纽约时报》2011年6月11日

维托里奥·格里高蒂，《建筑之内》，坎布里奇，哈佛大学设计学院，1996年

维托里奥·格雷高蒂，《区域的形态》，OASE建筑杂志总第80期，2010年

何深静，吴缚龙，《后改革开放时代中国的地产主导再开发：新天地项目案例分析》，《城市问题期刊》，第27期，2005年第1本，1–23页

迪克松·亨特，《景观建筑学历史中的园林与如画美学研究》，马萨诸塞州，麻省理工学院出版社，1992年

简·雅各布斯，《美国大城市的死与生》。兰登书屋，1961年

IMAGE CREDITS

弗雷德里克·詹明信，《当下的怀旧》，摘自《后现代主义，或，资本主义晚期的文化逻辑（后当代介入）》，达勒姆：杜克大学出版社，1991

布鲁斯·卡兹与朱莉·瓦格纳，《城市创新区的崛起》，《哈佛商业评论》，2014年11月12日

尼克·里昂，《吸引与连接：22@巴萨罗那创新区与巴塞罗那商业国际化》，《创新》杂志，第10期，2008，2-3页

刘亚晴，《上海："一城九镇"》，《那些年-瞭望东方周刊》2015-07-09

大卫·罗文塔尔，《制造遗产》，《历史和记忆》杂志 10.1（1998）

大卫·罗文塔尔，《古建保护的材料及其替代品》，《瞭望》杂志（1989年）

凯文·林奇与城市研究联合中心，《城市意象》，坎布里奇，马萨诸塞州：麻省理工学院出版社，1960年

奎多·莫泊格，维托里奥·格雷高蒂，《格雷高蒂国际建筑设计：城市景观的建筑》，纽约：里佐利出版社，2014年

莫森·莫斯塔法维，《生态城市主义》，坎布里奇，哈佛大学设计学院，2010年

中共上海市委，《上海市国民经济和社会发展第十三个五年规划纲要》，2016年1月29日

任雪飞，向过去前进：全球化上海的城市历史保护，《城市与社区》第7期，2008年3月，23-44页
保罗·吕滕，《创意，创新与城市发展》，《创意与城市：创意经济是如何转变城市的》(2005)

皮埃尔·莱克曼斯，《对待过去的中国式态度》第47卷，澳大利亚国立大学出版社，1986年

理查德·塞内特，《边界与界限》，载于里奇·博蒂特，德安·苏迪奇（主编），《生活于无尽城市中》，伦敦，2011

阿图罗·索里亚·普侬编，《塞尔达：整体城市化理论的五个基本点》，Electa出版社，1999年

迈克尔·斯托波，《区域的世界：经济全球化下的地域发展》，纽约：吉尔福德出版社，1997年

曼弗雷多·塔夫里，《维托里奥·格里高蒂：建筑与项目》，纽约，里佐利出版社，1982年

段义孚，《场所：一种实验性视角》，《地理研究》，1975年，65期（第2本）

城市土地学会，《城市土地学会案例分析：知识和创新社区》，华盛顿特区：城市土地学会，2015年

查尔斯·瓦德海姆，《景观城市主义读本》，纽约，普利斯顿建筑出版社，2006年

威尔海姆·米勒，《景观园林的草原精神》，阿姆赫斯特，马萨诸塞大学出版社，2002年

阿尔伯特·荣（2007），《上海新天地的场所提升与标志性》，《世界栖居》，2006年，第30期，245-260页

薛求理，周明浩，《引进与适应——上海"一城九镇"建设：维托里奥·格里高蒂的浦江新镇案例分析》，国际城市设计，2007年

周敏，《中国城：城市飞地的社会经济学潜力》，天普大学出版社，2010

Shanghai Aerial (6), the Former Residential Fabric before Demolishment (191), by Geza Radics on Flickr, https://www.flickr.com/photos/radicsge/8398746960/in/album-72157623433509439/

Shanghai Skyline 1990 (12), Shanghai Skyline 2010 (12), Cox, Savannah. The Incredible Evolution of Shanghai: 1990 to 1996 to 2010

Urbanized Area of Shanghai in 2005 (21), Urbanized Area of Shanghai in 2010 (21), Global Metropolitan Observatory, UC Berkeley

Continuous Monument Project by Superstudio (68), http://2.bp.blogspot.com/-yF6kJ9xKzZ8/U6txXBDROdI/AAAAAAAAEwc/MiVT4DjNH1M/s1600/superstudio-playa-1555x1020.jpg

A Greater West Park System, A Proposal for Integrating Parks and Gardens into the City Grid (69), Jensen, Jens, A greater west park system : after the plans of Jens Jensen, (Chicago: West Chicago Park Commissioners, 1920

Masterplan for Pujiang New Town (86), SWA Group, 2004

Masterplan for Pujiang New Town (86), Scacchetti Associati, 2004

Conceptual Masterplan for Pujiang New Town (87), Detailed Masterplan for Pujiang New Town (87), Gregotti Associati International, 2007

One City Nine Towns Plan (96), Hartog, H. Den. Shanghai New Towns : Searching for Community and Identity in a Sprawling Metropolis. Rotterdam: 010 Publishers, 2010.

Plan for Garden City (97), Howard, Ebenezer, Thomas, Ray, and Potter, Stephen. Garden Cities of Tomorrow. New Rev. ed. Eastbourne, East Sussex: Attic Books, 1985.

Wangjing Community, Beijing, China (113), https://antony2012.tuchong.com/12601300/

AVA High Line Mixed-Income Housing, New York, USA (115), http://observer.com/2014/06/ava-high-line-doesnt-want-your-security-deposit-she-just-wants-to-be-your-friend/

120 West 42nd Street, Manhattan, New York, USA as example of POPS (120), http://mdeas.com/wp-content/uploads/2014/08/03_Cubes_rev.jpg

No-Stop City. Archizoom, 1968-1970 (136), Plates from the project's publication in Domus, March 1971

The City of the Captive Globe. 1972 (137), Koolhaas, Rem. Delirious New York : A retroactive manifesto for Manhattan. New York: Oxford University Press, 1978.

Moema, São Paulo (140), http://vadebike.org/wp-content/uploads/2011/11/DSC07272.jpg

Inside a block in Eixample, Barcelona (143), http://you.ctrip.com/travels/barcelona381/1752136.html

Knowledge and Innovation Community Phase 2, University Avune Eastside Birdview (148), Knowledge and Innovation Community Phase 1, Jiangwan Stadium Westside Birdview (149), http://casestudies.uli.org/knowledge-and-innovation-community-shanghai/

KIC in Shanghai / Paris Rive Gauche / Facebook Campus in Silicon Valley, Before vs. After, (159), http://www.kic.net.cn/zh/learn, etc.

Xintiandi / Xintiandi Night View / M50 / M50 Night View / KIC Plaza / University Avenue (181), https://www.pinterest.com/pin/359936195195004289/, etc

Same Images of Chinese Cities (216), http://bbs.caup.net/read-htm-tid-34787-page-1.html

Old Factory in Hengmian (217), photographed by Guan Min

Wang Shu's Design in Zhejiang Province (222), http://www.wtoutiao.com/p/1578g6G.html

图像备注

上海全景鸟瞰（6），拆除前的居住区肌理（191），摄影：Geza Radics，https://www.flickr.com/photos/radicsge/8398746960/in/album-72157623433509439/

1990年上海的天际线（12），2010年上海的天际线（12），萨凡纳·考克斯，《上海的不可思议的演变：1990年至1996年至2010年》

2005年上海的城市化区域（21），2010年上海的城市化区域（21），加州大学伯克利分校，全球都会区域观测研究中心

"连续运动"项目，超级工作室（68），http://2.bp.blogspot.com/-yF6kJ9xKzZ8/U6txXBDROdI/AAAAAAAAEwc/MiVT4DjNH1M/s1600/superstudio-playa-1555x1020.jpg

大西部公园系统，将花园绿地融入城市格网的方案（69），延斯·延森，《大西部公园系统：在延斯·延森的方案之后》，芝加哥，西芝加哥公园管理局，1920年

浦江新镇规划设计详细图（86），斯卡切蒂设计事务所，2004

浦江新镇规划设计图（86），SWA集团，2004

浦江新镇规划设计概念图（87），浦江新镇规划设计详细图（87），格雷高蒂国际建筑设计，2007

一城九镇计划规划图（96），邓·哈尔托赫，《上海新城：追寻蔓延都市里的社区和身份》，鹿特丹：010出版社，2010

田园城市规划图（97），埃比尼泽·霍华德，雷·托马斯和史蒂芬·波特，《明日的田园城市》，伊斯特本，东苏塞克斯：阁楼出版社，1895

北京望京社区鸟瞰（113），https://antony2012.tuchong.com/12601300/

美国纽约曼哈顿高线AVA混合收入住宅（115），http://observer.com/2014/06/ava-high-line-doesnt-want-your-security-deposit-she-just-wants-to-be-your-friend/

美国纽约曼哈顿西42街120号，私人所有公共空间（120），http://mdeas.com/wp-content/uploads/2014/08/03_Cubes_rev.jpg

不停留城市，建筑视窗工作室，1968-1970年（136），刊于《Domus》杂志1971年三月期

囚禁中的城市，1972（137），摘自雷姆·库哈斯，《癫狂的纽约：一部曼哈顿的回溯性宣言》，纽约：牛津大学出版社，1978年，244页

圣保罗Moema街景（140），http://vadebike.org/wp-content/uploads/2011/11/DSC07272.jpg

巴塞罗那扩展区街区内部（143），http://you.ctrip.com/travels/barcelona381/1752136.html

创智天地二期，大学路东侧鸟瞰（148），创智天地一期，江湾体育场西侧鸟瞰（149），http://casestudies.uli.org/knowledge-and-innovation-community-shanghai/

上海创智天地 / 巴黎塞纳河左岸 / 硅谷Facebook总部开发前后（159），http://www.kic.net.cn/zh/learn 等

新天地 / 新天地夜景 / M50艺术区 / M50艺术区夜景 / 创智广场 / 大学路（181），https://www.pinterest.com/pin/359936195195004289/ 等

中国的"千城一面"现象（216），http://bbs.caup.net/read-htm-tid-34787-page-1.html

横沔老工厂（217），摄影：闵冠

建筑师王澍在浙江省的设计（222），http://www.wtoutiao.com/p/1578g6G.html